Editorial Policy

§ 1. Lecture Notes aim to report new developments - quickly, informally, and at a high level. The texts should be reasonably self-contained and rounded off. Thus they may, and often will, present not only results of the author but also related work by other people. Furthermore, the manuscripts should provide sufficient motivation, examples and applications. This clearly distinguishes Lecture Notes manuscripts from journal articles which normally are very concise. Articles intended for a journal but too long to be accepted by most journals, usually do not have this "lecture notes" character. For similar reasons it is unusual for Ph. D. theses to be accepted for the Lecture Notes series.

§ 2. Manuscripts or plans for Lecture Notes volumes should be submitted (preferably in duplicate) either to one of the series editors or to Springer- Verlag, Heidelberg . These proposals are then refereed. A final decision concerning publication can only be made on the basis of the complete manuscript, but a preliminary decision can often be based on partial information: a fairly detailed outline describing the planned contents of each chapter, and an indication of the estimated length, a bibliography, and one or two sample chapters - or a first draft of the manuscript. The editors will try to make the preliminary decision as definite as they can on the basis of the available information.

§ 3. Final manuscripts should preferably be in English. They should contain at least 100 pages of scientific text and should include
- a table of contents;
- an informative introduction, perhaps with some historical remarks: it should be accessible to a reader not particularly familiar with the topic treated;
- a subject index: as a rule this is genuinely helpful for the reader.

Further remarks and relevant addresses at the back of this book.

Lecture Notes in Mathematics   1680

Editors:
A. Dold, Heidelberg
F. Takens, Groningen

Springer

Berlin
Heidelberg
New York
Barcelona
Budapest
Hong Kong
London
Milan
Paris
Santa Clara
Singapore
Tokyo

Ke-Zheng Li   Frans Oort

Moduli of Supersingular Abelian Varieties

Springer

Authors

Ke-Zheng Li
Graduate School of Academia Sinica
Department of Mathematics
P.O. Box 3908
Beijing 100039, China
e-mail: kzli@math07.math.ac.cn

Frans Oort
Mathematisch Instituut
Budapestlaan 6
NL-3508 TA Utrecht, The Netherlands
e-mail: oort@math.ruu.nl

Cataloging-in-Publication Data applied for

Die Deutsche Bibliothek - CIP-Einheitsaufnahme

Li, Ke-Zheng:
Moduli of supersingular Abelian varieties / Ke-Zheng Li ; Frans
Oort. - Berlin ; Heidelberg ; New York ; Barcelona ; Budapest ; Hong
Kong ; London ; Milan ; Paris ; Santa Clara ; Singapore ; Tokyo :
Springer, 1998
  (Lecture notes in mathematics ; Vol. 1680)
  ISBN 3-540-63923-3

Mathematics Subject Classification (1991):
14K10, 14G15, 14L05, 14L15, 14D20, 14D22, 11G10, 11G15, 11R29

ISSN 0075-8434
ISBN 3-540-63923-3 Springer-Verlag Berlin Heidelberg New York

Typesetting: Camera-ready TeX output by the author
SPIN: 10553445   46/3143-543210 - Printed on acid-free paper

Contents

0. Introduction

0.1. Moduli of supersingular abelian varieties.

In this book we consider polarized abelian varieties over a field K of characteristic p. In positive characteristic abelian varieties naturally have an extra structure (not present in characteristic zero), given by properties of the subgroup scheme of "points" of p-power order.

Let us first explain briefly the main results, and then give more details. An elliptic curve E in characteristic p is called *supersingular* if it has no geometric points of order equal to p, i.e. let E be defined over a field $K \supset \mathbf{F}_p$, let k be an algebraically closed field containing K, then

$$E \text{ is supersingular} \overset{\text{def}}{\Longleftrightarrow} E[p](k) = 0$$

(some notation necessary to understand the contents of this introduction will be gathered together in 0.6 below). An abelian variety X in positive characteristic is called *supersingular* if it is geometrically isogenous to a product of supersingular elliptic curves, or in other words: if X is defined over $K \supset \mathbf{F}_p$, the dimension of X equals g, and k is an algebraically closed field containing K, then

$$X \text{ is supersingular} \overset{\text{def}}{\Longleftrightarrow} X \otimes k \sim E^g,$$

where E is a supersingular elliptic curve over k and "$\sim$" denotes isogeny equivalence (there are many other characterizations and properties, see 0.6 below).

Given $g \in \mathbf{Z}_{>0}$, a prime number p, and $d \in \mathbf{Z}_{>0}$, we denote by $\mathcal{A}_{g,d} \otimes \mathbf{F}_p$ the moduli space of all (X, λ), where X is an abelian variety of dimension g with a polarization λ of degree d^2, in characteristic p. We write

$$\mathcal{S}_{g,d} \subset \mathcal{A}_{g,d} \otimes \mathbf{F}_p \tag{0.1.1}$$

for the subset corresponding to all cases where X is supersingular, called the *supersingular locus* (in fact this is a closed algebraic subset).

One of the main results in this book is:

- *The dimension of $\mathcal{S}_{g,1}$ equals $[g^2/4]$* (the integral part of $g^2/4$), and

-

$$\#\{\text{irreducible components of } \mathcal{S}_{g,1}\} = \begin{cases} H_g(p, 1) & \text{if } g \text{ is odd,} \\ H_g(1, p) & \text{if } g \text{ is even.} \end{cases}$$

1

0.2. The supersingular locus in the moduli of abelian varieties.

A reader might wonder why these objects are studied, why they seem to be interesting.

In characteristic p the structure of the p-power torsion of an abelian variety is a canonical, extra structure (not present in this form in characteristic zero). Like abelian varieties can degenerate, also the "p-structure" can "degenerate". One can consider "ordinary abelian varieties" (those which have the maximal possible number of points of order p) as giving moduli points to the "interior" of the moduli space, while one could consider "degeneration" of the p-structure (while the abelian variety stays an abelian variety) as approaching to some kind of boundary. For example think of a variable elliptic curve in characteristic p, and put your fingers on the geometric points of order p; you feel that these come together when you specialize to a supersingular elliptic curve. This gives a fine structure on these moduli spaces which turn out to be of great help in understanding such moduli spaces (also the ones in characteristic zero). It turns out that the supersingular abelian varieties should be considered as the ones where the p-structure is most degenerated (analogous to the "cusps at infinity" in the boundary). For this reason these spaces $\mathcal{S}_{g,d}$ are interesting.

It turns out that the supersingular locus is quite different from other subsets defined by this kind of fine structures. The geometry of these spaces is very rich, it has certain geometric properties, but also a number theoretic flavor. Moreover, part of the structure of these spaces can be studied by purely algebraic methods which make some results even better accessible. We expect that while the supersingular locus is highly reducible (for g and p large), it might be true that the other loci (e.g. the ones given by Newton polygons which are not supersingular) can very well be irreducible. All this would give a strong approach to geometric, arithmetic and number theoretic study of moduli spaces of abelian varieties. For these reason we would like to understand the supersingular locus in itself very well. In this book we study various properties of these spaces. However, we leave aside how they are attached to the other interesting loci in the moduli space.

0.3. Polarizations, isogeny correspondences.

Why polarizations? We comment on some technical aspects of this work. First of all one should keep in mind that there is a difference between elliptic curves on the one hand (abelian varieties of dimension one), and abelian varieties of higher dimensions on the other hand. An abelian variety comes naturally with a rational point, the zero point. In the case of $g = 1$ this defines a divisor. For this reason every abelian variety of dimension one has a natural principal polarization; we can speak of moduli spaces of elliptic curves, meaning abelian varieties of dimension one with this natural (unique) principal polarization. However there are abelian varieties (of any dimension $g > 1$) which do *not* admit a principal polarization (this phenomenon occurs in all characteristics). And, when there is a principal polarization, it need not be unique (if $g > 1$); actually this is one of the main tools in the present study (we shall deal with all kind of mutually different principal polarizations on an abelian variety like E^g). When considering higher dimensions, it turns out that there is no good notion of moduli spaces of abelian varieties (without considering a polarization). For example, if one takes over the complex numbers the set of isomorphism classes of all abelian surfaces, there is no reasonable, natural geometric structure on this set (dividing out the equivalence relation, one might obtain non-Hausdorff spaces, this is a very classical topic, already known more than a century). Hence it is natural to consider *polarized* abelian varieties, when studying the cases with $g > 1$.

Isogenies. One can consider isogeny correspondences (Hecke correspondences) between components of moduli spaces. In characteristic zero, for an abelian variety X there are only a finite number of isogenies $X \dashrightarrow Y$ of a given degree, and this, in a certain way, simplifies the study of such correspondences. However in positive characteristic such correspondences (still well-defined) in general are not finite-to-finite; this accounts for several interesting and difficult aspects. For example it turns out that all of the components of the moduli space of polarized abelian varieties of dimension g over any field have the same dimension ($g(g+1)/2$ in fact), isogeny correspondences blow-up and down, but leave the dimension of the total spaces the same; miraculously the same holds for subsets defined by the p-rank. However components of the supersingular locus $\mathcal{S}_g = \bigcup_d \mathcal{S}_{g,d}$ can have different dimensions (when $g \geq 3$), in fact numbers between $[g^2/4]$ and $g(g-1)/2$ show up. This accounts for truly deep and difficult problems when studying certain closed subsets of the moduli space of polarized abelian varieties in positive characteristic. We shall deal with some of these questions.

0.4. PFTQs and parameter spaces of supersingular abelian varieties.

Flag type quotients. Here is the basic idea how to describe components of $\mathcal{S}_g$. Almost by definition, a supersingular abelian variety comes from E^g via an isogeny $E^g \to X$. This isogeny can be chosen to be purely inseparable, and in that case the group scheme $\ker(E^g \to X) \subset E^g$ is a repeated extension of the simple finite group scheme α_p. The basic idea is to describe the supersingular locus via all possibilities of such finite subgroup schemes of E^g. A hint what should be done is given by the fact that a "general" supersingular abelian variety canonically is the quotient of E^g via an isogeny of degree $p^{g(g-1)/2}$. Naturally this leads to the notion of *flag type quotients*.

For $g = 2$ a flag type quotient consists of

$$E^2 = X_1 \to E^2/\alpha_p \cong X_0. \tag{0.4.1}$$

Here we see that locally on the components of $\mathcal{S}_2$ the structure is given by varying α_p inside E^2, which is the same as giving a parameter on $\mathbf{P}^1$; in fact every component of $\mathcal{S}_2$ turns out to be a rational curve; the non-uniqueness of flag type quotients for some abelian surfaces gives rise to singularities of $\mathcal{S}_{2,1}$ which are transversal crossings of regular branches. For $g = 3$ a flag type quotient consists of

$$E^3 = X_2 \to E^3/(\alpha_p)^2 \cong X_1 \to X_1/\alpha_p \cong X_0. \tag{0.4.2}$$

3

In most cases, for a given X_0 such a sequence is unique, however for some cases it is not, and this causes the effect of singularities (in fact, of quite a bad type) on the components of $\mathcal{S}_{3,1}$. In general a flag type quotient, abbreviated by FTQ, for a supersingular abelian variety X_0 of dimension g is a sequence

$$E^g = X_{g-1} \to \cdots \to X_i/(\alpha_p)^i \cong X_{i-1} \to \cdots \to X_1 \to X_0. \qquad (0.4.3)$$

where E is a supersingular elliptic curve (which, for $g \geq 2$ can be chosen once and for all).

Polarized flag type quotients. In order to obtain components of $\mathcal{S}_g$ we have to put a polarization on the X_0 in consideration. This can be done by choosing a polarization on E^g, of degree $d^2 \cdot p^{g(g-1)}$ in fact, which descends via the flag type quotient (0.4.3) to a polarization λ_0 of degree d^2 on X_0. Then in (0.4.3) each X_i is equipped with a polarization of the appropriate degree, such that they form a descending chain of polarizations. Such a sequence is called a *polarized flag type quotient*, abbreviated $PFTQ$, for (X_0, λ_0). The moduli space P of PFTQs exists, and there is a surjective morphism

$$\Psi : P \twoheadrightarrow \mathcal{S}_g \subset \mathcal{A}_g. \qquad (0.4.4)$$

As we have already mentioned earlier, also for the polarized case, for "general" polarized supersingular abelian varieties a PFTQ is unique. However the phenomenon that for special cases it is not unique causes that the morphism Ψ is blowing down, for $g \geq 3$, certain subsets of this parameter space P to subsets of $\mathcal{S}_g$. For $g \geq 4$ this turns out to be rather bad: it might blow down a whole component of P to a proper closed subset of a component of $\mathcal{S}_g$ (this even happens above $\mathcal{S}_{g,1}$): we call such a component of P a "garbage component". The existence of these was for a long time the obstacle to describe all components of $\mathcal{S}_{g,1}$ in the case when $g \geq 4$. One of the main points of the present work is the (rather technical) definition of a "rigid PFTQ". This notion singles out a Zariski open subset $P' \subset P$, on which the map

$$\Psi : P' \to \mathcal{S}_g \qquad (0.4.5)$$

is indeed *finite to one and surjective* (for example the closure of P' in P does not contain any of the garbage components). For a general principally polarized supersingular abelian variety its (canonical) polarized flag type quotient is automatically rigid. As is usual in moduli theory, once a good moduli-theoretic description is given, one can proceed. In fact we show that any component of P' maps finite to one onto a (non-empty open set of a) component of $\mathcal{S}_g$, and any component of the latter one can be obtained in precisely one way along this line. Once we have arrived this point, it is clear how to proceed:

- *compute the dimension of every component of P'; this turns out to be equal to $[g^2/4]$ above $\mathcal{S}_{g,1}$, and*

- *compute the number of isomorphism classes of polarizations with the required properties on E^g in order to describe the number of components of $\mathcal{S}_{g,d}$.*

- For $d = 1$, the case of principally polarized supersingular abelian varieties, this amounts to considering all polarizations

$$\eta : E^g \to (E^g)^t, \quad \text{with} \quad \ker(\eta) = E^g[F^{g-1}],$$

4

where F is the Frobenius morphism of E^g.

- For g odd this amounts to the same as the number of equivalence classes of principal polarizations on E^g, which is known to be equal to the class number $H_g(p,1)$, hence this is the number of geometrically irreducible components of $\mathcal{S}_{g,1}$. For g even we obtain the class number $H_g(1,p)$ as the number of components.

0.5. Strategy for proving the main properties of the moduli of rigid PFTQs.

It seems natural to consider for a given g and for a given $0 \leq m \leq g-1$ a moduli space $\mathcal{V}_m$ of polarized, rigid, partial flags

$$E^g \cong Y_{g-1} \to \cdots \to Y_{m+1} \to Y_m. \qquad (0.5.1)$$

In this way we obtain a sequence of spaces and "truncation maps"

$$\mathcal{V}_0 \to \cdots \to \mathcal{V}_{g-2} \to \mathcal{V}_{g-1} = \{\text{one point}\}, \qquad (0.5.2)$$

where $\mathcal{V}_m$ ($0 \leq m \leq g-1$) is the moduli of sequences (0.5.1), and in particular $\mathcal{V}_0$ is the moduli of PFTQs.

This idea has to be refined. The heart of the proof of the main result on $\mathcal{S}_{g,1}$ uses complete induction from $g-2$ to g, by constructing moduli spaces which combine a partial flag $Y_{g-1} \to \cdots \to Y_m$ for genus g, with a complete flag for genus $g-2$, related in some way, with extra properties, which ensure that the incomplete flag can be completed; this is a tricky condition (see condition d) in 11.3). For example consider the explicit condition for the case $g = 4$ (see 9.7). The moduli spaces thus obtained fit into a sequence of morphisms; each turns out to be a smooth epimorphism of relative dimension one. Once this is proved the main result follows.

One may note the different behavior between $\mathcal{S}_{g,1}$ for odd g on the one hand, and the same for even g on the other hand. This is reflected in the fact that we consider polarizations of E^g whose kernel is $E[F^{g-1}]$. For

$$g - 1 = 2m \quad \text{we have} \quad E[F^{g-1}] = E^g[p^m],$$

and such a polarization equals p^m times a principal polarization. For

$$g - 1 = 1 + 2m \quad \text{we have} \quad E[F^{g-1}] = E^g[p^m F],$$

which gives polarizations with a different type of behavior. Also the difference is found back in the proof (which works by induction from $g-2$ to g).

There is one more technical point we would like to mention. Let G be the formal group (in this case also the p-divisible group) of a supersingular elliptic curve E. A principal polarization on E^g gives a quasi-polarization on G^g. It turns out that any two principal polarizations on E^g give equivalent quasi-polarizations on G^g (a kind of global-local property). This simplifies the description of the components of

5

the moduli of rigid PFTQs. and the class numbers involved describe the number of components of $S_{g,1}$.

For further discussions of technical points, of examples, of other results, we refer to the main text.

0.6. Some definitions used in the introduction.

In this section we collect some definitions and explain some of the terminology used in the introduction. We shall write K for a field and k for an algebraically closed field, we usually assume $K \subset k$. A term like "geometrically integral" will mean "integral after $\otimes_K k$".

Definition: An *abelian variety* defined over K is a complete group variety over K. i.e. a K-group scheme which is proper over K and geometrically integral. Note that for an abelian variety the group law is commutative.

An *abelian scheme* $X \to S$ is an S-group scheme which is smooth and proper over S such that all fibres are abelian varieties. (For general references see [53] or [56].)

Definition: An *elliptic curve* over a field K is an abelian variety over K of dimension one.

Note that the following properties are equivalent:

i) E is an elliptic curve over K, isomorphisms are isomorphisms of group varieties.

ii) E is an elliptic curve over K, isomorphisms are isomorphisms of varieties, preserving the point zero.

iii) E is a an algebraic curve smooth and proper over K, geometrically connected, of genus 1, with a given K-rational point $0 \in E(K)$; isomorphisms are isomorphisms of algebraic curves, preserving the point 0.

iv) $E \subset \mathbf{P}_K^2$ is a projective, plane curve of degree 3, smooth over K, with a given point $0 \in E(K)$; isomorphisms are given by projective isomorphisms preserving the point 0.

v) $E \subset \mathbf{P}_K^2$ is given by the equation

$$Y^2 Z + a_1 XYZ + a_3 YZ^2 = X^3 + a_2 X^2 Z + a_4 XZ^2 + a_6 Z^3$$

with $a_i \in K$, with the discriminant non-zero (we do not write it down here). where the point zero on the curve is given by $(0 : 1 : 0)$; isomorphisms are given by projective isomorphisms preserving the point zero.

For an additive abelian group A and an integer n we write $A[n]$ for the kernel of $\times n : A \to A$. For a commutative group scheme G we write

$$G[n] = \ker(n \cdot \mathrm{id}_G : G \to G).$$

considered as a subgroup scheme of G. Note that we have

$$G[n](k) = G(k)[n]$$

6

if G is a group scheme over some subfield of k.

If G is a group scheme over a field K of characteristic $p > 0$, we have the *relative Frobenius homomorphism* $F_{G/K} : G \to G^{(p)}$ (see 2.3), and we write $G[F] = \ker(F_{G/K})$.

A finite surjective homomorphism $X \to Y$ between abelian schemes over S is called an *isogeny*. If there exists an isogeny $\phi : X \to Y$, then we say X is *isogenous* to Y, denoted by $X \sim Y$. It turns out that $\sim$ is an equivalence relation: since $n = \deg(\phi)$ annihilates $\ker(\phi)$, we have $\ker(\phi) \subset X[n]$, hence there exists an isogeny $\psi : Y \to X$ such that $\psi \circ \phi = n \cdot \mathrm{id}_X$, in particular $Y \sim X$.

For an abelian scheme $X \to S$ there is a *dual* abelian scheme $X^t \to S$. A *polarization* on $X \to S$ is by definition an S-isogeny $\lambda : X \to X^t$ which on every geometric fibre is given by an ample divisor (see [56, Definition 6.3]). A polarization is called *principal* if it is an isomorphism.

Here is a way to construct abelian varieties which do not admit a principal polarization. Choose an integer $g \in \mathbf{Z}_{\geq 2}$, choose a field k and an abelian variety X over k such that $\mathrm{End}(X) \cong \mathbf{Z}$ (for every characteristic such abelian varieties exist). If X does not admit a principal polarization we are done. If X does admit a principal polarization, choose an integer $n \in \mathbf{Z}_{\geq 2}$ prime to $\mathrm{char}(k)$, and a cyclic subgroup $N \subset X$ of order n; then $Y := X/N$ does not admit a principal polarization. This can be seen as follows: there does exist an isogeny $Y^t \to X^t$ with kernel cyclic of order n, a principal polarization on Y would give

$$h := (X \to X/N = Y \cong Y^t \to Y^t/N = X^t \cong X) \in \mathrm{End}(X) \cong \mathbf{Z};$$

if $h = m \cdot \mathrm{id}_X$, then on the one hand $\ker(h) \cong (\mathbf{Z}/m\mathbf{Z})^{2g}$; on the other hand $\ker(h)$ is an extension of $\mathbf{Z}/n\mathbf{Z}$ by $\mathbf{Z}/n\mathbf{Z}$, and we obtain a contradiction with $g \geq 2$.

In positive characteristic one easily constructs examples (even over $\bar{\mathbf{F}}_p$) of abelian varieties which do not admit a separable polarization.

For the definition and existence of the moduli spaces

$$\mathcal{A}_{g,d,n} \to \mathrm{Spec}(\mathbf{Z}[1/n])$$

we refer to [56, 7.2].

Definition: Let X be an abelian variety over $K \supset \mathbf{F}_p$. Then there exists an integer $f = f(X)$, called the *p-rank* of X, such that

$$X[p](k) \cong (\mathbf{Z}/p\mathbf{Z})^f.$$

Note that $0 \leq f \leq g = \dim(X)$. An abelian variety X is called *ordinary* if its p-rank is maximal, i.e.

$$X \text{ is ordinary} \iff X[p](k) \cong (\mathbf{Z}/p\mathbf{Z})^{\dim(X)}.$$

If X and Y are isogenous to each other then $f(X) = f(Y)$.

An elliptic curve in positive characteristic is called *supersingular* if it is not ordinary, i.e. if it does not contain a point of order p.

Over an algebraically closed field $k \supset F_p$ there are, up to isomorphism, exactly three group schemes whose structure ring has rank p over k:

$$\mu_p, \qquad \alpha_p, \qquad Z/pZ.$$

Moreover when G is a finite group scheme over K then $G \cong \alpha_p$ if and only if $G \otimes k \cong \alpha_p$.

An ordinary abelian variety X in characteristic p is characterized by:

$$X \text{ is ordinary} \iff X[p] \otimes k \cong (\mu_p)^{\dim(X)} \times (Z/pZ)^{\dim(X)}.$$

We see that an elliptic curve E is supersingular (when defined over a field of characteristic p) if and only if $E[F] \cong \alpha_p$.

Complex multiplications and supersingularity. We give some more information, which might explain the terminology "supersingular".

Let X be an abelian variety of dimension g. If $\mathrm{End}^0(X) = \mathrm{End}(X) \otimes Q$ contains a commutative semi-simple algebra of rank $2g$ over Q, then we say X has "*sufficiently many complex multiplications (smCM)*".

Let us explain first the case of elliptic curves. In characteristic zero, say over the field C of complex numbers, an elliptic curve E either has the property $\mathrm{End}(E) = Z$ (and we say "E has no complex multiplications"), or $Z \subset \mathrm{End}(E)$ is a proper inclusion (and we say "E has complex multiplications", or CM in short). It is known classically that an elliptic curve with CM can be defined over a finite extension of Q (i.e. over a number field); classically the j-invariant in this case was called a "singular j-invariant"; in this case $\mathrm{End}(E)$ is an order in an imaginary quadratic field.

In positive characteristic there are more possibilities. Suppose $F_p \subset K$, let k be an algebraically closed field containing K, and let E be an elliptic curve over K. Then one of the following three properties holds:

i) $\mathrm{End}(E \otimes k) = Z$. In this case E cannot be defined over a finite field. Equivalently: its j-invariant is transcendental over F_p. In this case E is ordinary.

ii) $\mathrm{End}(E \otimes k)$ is an algebra of rank 2 over Z. In this case $\mathrm{End}(E)$ is an order in an imaginary quadratic field in which p is split and E (or its j-value) is called *singular*. Also in this case E is ordinary, and it can be defined over a finite field.

iii) $\mathrm{End}(E \otimes k)$ is an algebra of rank 4 over Z. In this case its endomorphism algebra $\mathrm{End}^0(E \otimes k) := \mathrm{End}(E \otimes k) \otimes Q$ is a central simple algebra of degree 4 over Q ramified exactly at ∞ and at p. Moreover $E[p](k) = 0$, and $j(E) \in F_{p^2}$, and E (or its j-value) is called *supersingular*.

A little warning: it might happen for an elliptic curve E defined over a field K that $\mathrm{End}(E)$ has rank two over Z, and $\mathrm{End}(E \otimes k)$ has rank four over Z. In that case, $\mathrm{char}(K) = p > 0$, and $\mathrm{End}(E)$ is an order in an imaginary quadratic number field in which p does not split.

For every p there exists a supersingular elliptic curve over $\mathbf{F}_p$. The number h_p of isomorphism classes of supersingular elliptic curves (over k, say over $\bar{\mathbf{F}}_p$) is finite, this number is a classical invariant (we will come back to this, see 9.1). Any two supersingular elliptic curves over $\bar{\mathbf{F}}_p$ are isogenous to each other.

Supersingular abelian varieties. For abelian varieties of arbitrary dimension over $K \supset \mathbf{F}_p$ there are many possibilities for the structure of the p-torsion subgroup scheme, and for $\mathrm{End}(X)$. If $\mathrm{End}(X)$ is larger than $\mathbf{Z}$ one could say "X has complex multiplications" (but in general we *don't*), the rank of $\mathrm{End}(X)$ over $\mathbf{Z}$ can have several values. An abelian variety X of dimension g is supersingular if and only if $\mathrm{End}(X \otimes k)$ is of rank $(2g)^2$ over $\mathbf{Z}$.

We remark that in higher dimension (over $K \supset \mathbf{F}_p$) there are showing up some complexities (or, if you like, some extra interesting features), not present for elliptic curves:

- If $\dim(X) \leq 2$ and $X[p](k) = 0$ then X is supersingular, however,
- for every $g \geq 3$ there is an abelian variety X of dimension g with $X[p](k) = 0$ which is not supersingular.
- As Tate showed (see [97]), any abelian variety X defined over a finite field has sufficiently many complex multiplications. However the converse is not quite true, as Grothendieck showed (see [66]): an abelian variety X with smCM is isogenous to an abelian variety defined over a finite extension of the prime field (but X need not be defined over a finite field in the case of finite characteristic, see Appendix A.3 for more details). In particular:
- For every $g \geq 2$ there exist positive dimensional non-trivial families of supersingular abelian varieties, in other words: in these cases there exist supersingular abelian varieties not defined over a finite field, but they are isogenous to an abelian variety defined over a finite field, e.g. to E^g, where E is a supersingular elliptic curve .
- In fact, for integers $g \geq 2$ and f with $0 \leq f \leq g - 2$ there exist abelian varieties in characteristic $p > 0$ of dimension g, with p-rank equal to f, which have smCM, but which cannot be defined over a finite field.

Newton polygons (not used in this book). Using Dieudonné-Manin theory one can define for every abelian variety (of dimension g) in characteristic p its *Newton polygon* (NP). This is a polygon which starts at $(0,0)$, ends at $(2g, g)$, which is lower convex, and has break points with integral coordinates. Moreover for the slope λ of every side of this polygon we have $0 \leq \lambda \leq 1$. In fact ordinary abelian varieties are characterized by the fact that the NP has g slopes equal to 0, and g slopes equal to 1. Supersingular abelian varieties turn out to be characterized by the fact that all $2g$ slopes are equal to $1/2$. Any other of these Newton Polygons is between these two. We see that from this point of view the ordinary abelian varieties are the most general ones, and the supersingular ones, studied in this book, are the most particular ones.

N.B. *We should mention results previously obtained, we should acknowledge contributions to this topic made in the past. In order not to overburden this short* introduction, this will be done in the Appendix of this book, where we give a historical survey of (part of) this topic.

Convention. In the text we use the section numbers to index definitions, theorems, remarks etc., for example Lemma 6.1 means the lemma in 6.1.

Acknowledgements.

We wish to thank especially T. Ekedahl, K. Feng, T. Ibukiyama, A.J. de Jong, T. Katsura, W.C. Winnie Li, A. Ogus, and many other colleagues for valuable discussions, suggestions, and their patience in listening to our supersingular expositions.

The first author wishes to thank the constant support of National Science Foundation Committee (NSFC) of China and the support of Nederlandse Organisatie voor Wetenschappelijk Onderzoek (NWO) of The Netherlands, also wishes to thank the University of Utrecht for the hospitality and the stimulating environment of mathematical research. In 1991 the first author was invited by Prof. T. Oda and Prof. M. Miyanishi to visit Japan, and gave a special talk on the result of this text in the annual meeting of the Mathematical Society of Japan (MSJ). This enabled him to discuss with several specialists on this topic. He also wishes to thank MSJ, Tohoku University and Osaka University for their support and hospitality.

The second author wishes to thank T. Oda and T. Katsura for many years of growing results on this topic.

The main material for this book originated several years ago, and we feel we should apologize that the publication was somewhat slow.

1. Supersingular abelian varieties

Throughout this book we will denote by p a prime number, which is fixed unless otherwise specified. We denote by K a field of characteristic p and by k an algebraically closed field containing K. For any K-scheme X and any field extension $K' \supset K$, we will denote $X \times_{\mathrm{Spec}(K)} \mathrm{Spec}(K')$ simply by $X \otimes_K K'$, or even $X \otimes K'$ if there is no confusion.

In this chapter every scheme is defined over some K, unless otherwise specified.

1.1. Supersingular elliptic curves.

Let E be an elliptic curve over K. Then either E has a geometric point of order exactly p and E is called *ordinary*, or E has no geometric point of order p and E is called *supersingular*.

It is well known that the number of isomorphism classes of supersingular elliptic curves over k is finite (the number is roughly $p/12$ and $\leq [p/12]+2$, cf. [23, Corollary IV.4.23], see (9.1.4) for an exact formula), and any two supersingular elliptic curves over k are isogenous (i.e., there exists a finite-to-one morphism from one to the other, cf. [9, p. 252]).

1.2. Endomorphism algebra of supersingular elliptic curves.

For every prime number p there exists an elliptic curve E over the prime field F_p such that its relative Frobenius

$$F : E \to E^{(p)} \cong E \tag{1.2.1}$$

satisfies

$$F^2 + p = 0 \tag{1.2.2}$$

(cf. [97, pp. 139-140], [98, p. 96], [100, Theorem 4.1.5]). For the rest of this book we fix a choice of such an E over F_p for each p. Note that E has the property

$$\mathrm{rank}_{\mathsf{Z}}(\mathrm{End}(E)) = 2 \tag{1.2.3}$$

and

$$\mathcal{O} := \mathrm{End}(E \otimes \mathsf{F}_{p^2}) = \mathrm{End}(E \otimes k) \tag{1.2.4}$$

has rank 4 over Z; it is a maximal order in the quaternion algebra

$$B := \mathrm{End}^0(E \otimes k) = \mathrm{End}(E \otimes k) \otimes \mathsf{Q} \cong Q_{\infty,p} \tag{1.2.5}$$

which is split at every prime number $l \neq p$ (see [9, p.199]).

11

1.3. p-divisible groups and duality.

Fix a base scheme S. For a commutative group scheme $\pi : G \to S$ and any positive integer n, we will denote

$$G[n] := \ker(n_G : G \to G) \tag{1.3.1}$$

where $n_G = n \cdot \mathrm{id}_G$ is the multiplication by n.

If π is flat and finite, we denote by G^D the Cartier dual of G over S, i.e. the structure O_S-algebra of G^D is isomorphic to $\mathcal{H}om_{O_S}(\pi_* O_G, O_S)$, the dual Hopf algebra of $\pi_* O_G$ over O_S (cf. e.g. [64, I.2]).

Let $\mathfrak{C}_S^1$ be the category of flat finite commutative group schemes over S whose ranks are powers of p. Let $\mathfrak{C}_S$ be the category of formal inductive limits in $\mathfrak{C}_S^1$:

$$G = \varinjlim_n G_n \tag{1.3.2}$$

satisfying

   a) $G_n = G_{n+1}[p^n]$ for each n.

Such a G is called a *commutative formal group*, and it is called a *p-divisible group* if in addition that

   b) p_G is an epimorphism.

Condition b) is equivalent to that the homomorphism $G_{n+1} \to G_n$ induced by p_G is an epimorphism for each n, in this case we have induced monomorphisms $G_n^D \hookrightarrow G_{n+1}^D$ and we denote

$$G^t = \varinjlim_n G_n^D, \tag{1.3.3}$$

called the *Serre dual* of G.

An *isogeny* of p-divisible groups is an epimorphism with finite kernel. If there is an isogeny from G to G', then we say G and G' are *isogenous* to each other, denoted by $G \sim G'$.

1.4. The formal isogeny type of an abelian variety.

For an abelian scheme X over S we define:

$$\varphi_p X := \varinjlim_i (X[p^i]). \tag{1.4.1}$$

(Sometimes this is denoted by $X[p^\infty]$.) This is a p-divisible group, called the *Barsotti-Tate group* of X.

Denote by X^t the dual abelian scheme of X. Clearly we have $\varphi_p(X^t) \cong (\varphi_p X)^t$ (see (1.3.3) and [56, III.15]).

Over an algebraically closed field k, the p-divisible groups have been classified up to isogeny by the Dieudonné-Manin theory. For the case of Barsotti-Tate groups we have:

$$\varphi_p\, X \ \sim\ \sum_i (G_{m_i, n_i} \oplus G_{n_i, m_i}) \bigoplus G_{1,1}^{\oplus s} \bigoplus (G_{1,0} \oplus G_{0,1})^{\oplus f} \tag{1.4.2}$$

$$(m_i > n_i > 0,\ \text{g.c.d.}(m_i, n_i) = 1);$$

here $G_{m,n}$ is a simple p-divisible group over k (see [48, p.37]) which has the following properties: $\dim_k \text{Lie}(G_{m,n}) = m$ and $\dim_k \text{Lie}(G^t_{m,n}) = n$ (cf. (2.1)). Such a decomposition (up to isogeny) is called a *formal isogeny type*. (The symmetry in (1.4.2) is called the "Manin symmetry condition".) We say that this formal isogeny type is

$$\textbf{supersingular} \text{ iff } \varphi_p X \sim G^{\oplus g}_{1,1}$$

where $g = \dim(X)$. In general, an abelian variety X over K is called *supersingular* if $\varphi_p(X \otimes_K k)$ is supersingular.

Convention: The p-divisible group $G_{m,n}$ is defined over F_p, however for any field extension $\mathsf{F}_p \subset K$ we shall write $G_{m,n}$ instead of $G_{m,n} \otimes K$. The same for α_p, μ_p, G_a, G_m, in case no confusion can arise.

1.5. The a-number.

For a commutative group scheme X over a field K we define

$$a(X) = \dim_K \text{Hom}(\alpha_p, X). \tag{1.5.1}$$

If $K \subset K'$ then

$$\dim_K \text{Hom}(\alpha_p, X) = \dim_{K'} \text{Hom}(\alpha_p, X \otimes K'), \tag{1.5.2}$$

i.e. the a-number does not depend on the field we are working over. Furthermore, there is a smallest subgroup scheme $A(X) \subset X$ containing all of the images of $\alpha_p \to X$; note that its rank is $p^{a(X)}$ (see 2.5).

1.6. A characterization of supersingularity.

Supersingular abelian varieties are distinguished from other abelian varieties by the following property. For any formal isogeny type which is not supersingular, there exists a simple abelian variety over k having this formal isogeny type (cf. [44, p.47]). However for supersingular formal isogeny types with $g \geq 2$ the situation is different:

Fact. *Let X be an abelian variety of dimension $g \geq 2$. Then*
   i) X is supersingular if and only if $X \otimes k \sim E^g \otimes k$, thus:

$$\varphi_p(X \otimes k) \sim G^g_{1,1} \iff X \otimes k \sim E^g \otimes k;$$

   ii) $$a(X) = g \iff X \otimes k \cong E^g \otimes k.$$

The first statement can be found in [67, Theorem 4.2]. For the second statement one uses [69, Theorem 2]: we see that $a(X) = g$ iff $X \otimes k$ is isomorphic to a product $E_1 \times ... \times E_g$ of supersingular elliptic curves over k; by a theorem due to Deligne (using a calculation by Eichler) and to Ogus (cf. [62, Theorem 6.2] and [95, Theorem 3.5]), we know that for any $g \geq 2$ and any supersingular elliptic curves $E_1, ..., E_{2g}$ over k,

$$E_1 \times ... \times E_g \cong E_{g+1} \times ... \times E_{2g}. \tag{1.6.1}$$

Hence (for $g \geq 2$):

$$a(X) = g \iff a(X \otimes k) = g \iff X \otimes k \cong E^g \otimes k. \tag{1.6.2}$$

1.7. Superspecial and supergeneral abelian varieties.

Definition. We say that

$$X \text{ is } \textbf{superspecial} \text{ if } a(X) = \dim(X)$$

(and by Fact 1.6 this is equivalent to: $X \otimes k \cong E^g \otimes k$). We say that X is *supergeneral* if X is supersingular and $a(X) = 1$.

Note that any abelian subvariety of a superspecial abelian variety X is also superspecial. If X is superspecial and $G \subset X$ is a finite étale subgroup scheme, then X/G is also superspecial.

1.8. Minimal isogenies.

Lemma. *Let $g \geq 2$ and let X be a supersingular abelian variety of dimension g over K. Then there is a superspecial abelian variety Y over K and a K-isogeny (called a minimal isogeny)*

$$\rho : Y \to X \tag{1.8.1}$$

such that for any superspecial abelian variety Y' over K and any homomorphism $\rho' : Y' \to X$, there is a unique homomorphism $\phi : Y' \to Y$ such that $\rho' = \rho \circ \phi$. (Hence the minimal isogeny is unique up to isomorphism.)

Dually, there exists a superspecial abelian variety Z over K and a K-isogeny $\gamma : X \to Z$ such that for any superspecial abelian variety Z' over K and any homomorphism $\gamma' : X \to Z'$, there is a unique homomorphism $\psi : Z \to Z'$ such that $\gamma' = \psi \circ \gamma$.

Proof. Define inductively $X_0 = X$ and $X_i = X_{i-1}/A(X_{i-1})$ (see 1.5 above). Then $a(X_i) > a(X_{i-1})$ unless $a(X_{i-1}) = g$ (see [45, p.337]). In particular, X_{g-1} is superspecial. Let $Y = X_{g-1}^{(p^{g-1})}$. By induction we see that $\ker(X_{g-1}^t \to X^t)$ is killed by F^{g-1}. Hence $F_{X_{g-1}^t/K}^{g-1}$ factors through X^t and we get an induced homomorphism $\rho : Y \to X$.

Let Y' be a superspecial abelian variety over K and $\rho' : Y' \to X$ be a homomorphism. Let $Y_i' = Y'^{(p^i)}$ $(0 \leq i \leq g-1)$. We see ρ' (and $F_{Y'/K}^i$) induces $Y_i' \to X_i$ $(0 \leq i \leq g-1)$ by induction. In particular, we get an induced homomorphism $Y_{g-1}' \to X_{g-1}$. Since $\ker(Y \to X_{g-1}) = Y[F^{g-1}]$, we have an induced homomorphism $\phi : Y' \to Y$ such that $\rho' = \rho \circ \phi$, which is obviously unique.

The remaining statements come from duality. Q.E.D.

We denote $\{Y, \rho\}$ (resp. $\{Z, \gamma\}$) simply by $S^0 X$ (resp. $S_0 X$). See 5.6 and [48, p.38] for the corresponding notation on Dieudonné modules.

14

Remark. From the proof of Lemma 1.8 we see that $\ker(\rho)$ is infinitesimal (cf. 2.3), hence $\deg(\rho) = p^n$, where $n = \sum_{i=0}^{g-2} g - a(X_i)$. By induction we also see that $a(X_i) \geq i+1$, hence $n \leq g(g-1)/2$. It was shown in [61, Theorem 2.2] that when X is *supergeneral* we have $a(X_i) = i+1$, hence $n = g(g-1)/2$. Clearly $n < g(g-1)/2$ when $a(X) > 1$ because $a(X_i) \geq i+2$ in this case.

1.9. The supersingular locus.

Let $\mathcal{A}_{g,d,n}$ be the moduli scheme of abelian varieties together with a polarization of degree d^2 and a level n-structure (which is a fine moduli scheme when $n \geq 3$). In particular $\mathcal{A}_{g,d} = \mathcal{A}_{g,d,1}$ is the coarse moduli scheme of abelian varieties together with a polarization of degree d^2. (For the definition and construction of $\mathcal{A}_{g,d,n}$, see [53] or [56].)

For each prime number $p \nmid n$, the points in $\mathcal{A}_{g,d,n} \otimes \mathbb{F}_p$ representing supersingular abelian varieties form a (Zariski) closed subset ([38, Corollary 2.3.2]), whose reduced induced structure is called the *supersingular locus* in $\mathcal{A}_{g,d,n}$, denoted by $\mathcal{S}_{g,d,n}$. We also denote $\mathcal{S}_{g,d} = \mathcal{S}_{g,d,1}$. In particular $\mathcal{S}_{g,1}$ is the supersingular locus in the coarse moduli $\mathcal{A}_{g,1}$ of principally polarized abelian varieties.

In Chapter 13 we will define the supersingular locus $\mathcal{S}_{g,1,n}$ as a moduli scheme (see Corollary 13.14).

For any irreducible subscheme $V \subset \mathcal{A}_{g,d} \otimes \mathbb{F}_p$, we denote by $a(-/V)$ the a-number of the abelian variety represented by the generic point of V.

2. Some prerequisites about group schemes

In this chapter S denotes a base scheme.

2.1. Some notions about group schemes.

For a (separated) group scheme $\pi : G \to S$, we will denote by $m : G \times_S G \to G$ the multiplication, $o : S \to G$ the unit section, and $\iota : G \to G$ the inverse. Let $\mathcal{M} = \ker(o^*)$ be the ideal sheaf of the closed immersion o. Then $\omega_{G/S} = o^* \mathcal{M}$ is called the sheaf of (left) invariant differentials. We have

$$\Omega^1_{G/S} \cong \pi^* \omega_{G/S}, \quad \mathrm{Lie}(G/S) \cong \mathcal{H}om_{O_S}(\omega_{G/S}, O_S), \tag{2.1.1}$$

where O_S is the structure sheaf of S and $\mathrm{Lie}(G/S)$ is the sheaf of (left) invariant Lie algebras of G over S.

If π is finite and $\pi_* O_G$ is locally free of rank r as an O_S-module, then we simply say that G has *rank* r.

2.2. Universal finite subgroup scheme.

We will use the following ([45, Lemma 2.8])

Fact. *Let G be a finite group scheme over S. Then for any positive integer r, the functor*

$$((S\text{-schemes})) \longrightarrow ((\text{sets}))$$
$$T \mapsto \{\text{closed subgroup schemes of } G \times_S T, \text{ flat of rank } r \text{ over } T\}$$

is represented by a relatively projective scheme over S.

2.3. Frobenius and Verschiebung.

In the following we assume S is an $\mathbb{F}_p$-scheme.

The *absolute Frobenius* of S is the morphism $F_S : S \to S$ which sends every point of S to itself, but F_S^* sends every section of O_S to its pth power. For any S-scheme $\pi : X \to S$, obviously we have $F_S \circ \pi = \pi \circ F_X : X \to S$. For any $n > 0$, denote by $X^{(p^n)}$ the pull-back of π and F_S^n, then we have an induced S-morphism $F_{X/S}^n : X \to X^{(p^n)}$ such that $q \circ F_{X/S}^n = F_X^n$, where $q : X^{(p^n)} \to X$ is the projection. (Warning: F_X is usually *not* an S-morphism!) We call $F_{X/S} = F_{X/S}^1$ the *relative Frobenius* of X over S.

Note that $X^{(p)}$ is functorial with respect to X (i.e. $X \mapsto X^{(p)}$ gives a functor j from $((S\text{-schemes}))$ to itself), and $F_{X/S}$ is a natural S-morphism (i.e. it gives a natural transformation from $\mathrm{id}_{((S\text{-schemes}))}$ to j). In particular, if X is a group scheme over S, so is $X^{(p)}$, and $F_{X/S}$ is a canonical homomorphism of group schemes over S.

16

Let G be a group scheme over S. For convenience, we denote

$$G[F^n] := \ker(F_{G/S}^n : G \to G^{(p^n)}).$$ (2.3.1)

We say G is *infinitesimal* (or *local*) if $\mathcal{M}$ is contained in the nilradical of O_G. When π is of finite type, this is equivalent to: $F_{G/S}^n = 0$ for some n.

If G is flat and commutative, then there is an S-homomorphism $V_{G/S} : G^{(p)} \to G$ such that $V_{G/S} \circ F_{G/S} = p_G$ and $F_{G/S} \circ V_{G/S} = p_{G^{(p)}}$ (see [8, I, pp. 440-442]). In particular, if $G \to S$ is finite, one can take $V_{G/S}$ to be the dual of $F_{G^D/S}$, where G^D is the Cartier dual of G (cf. [6, §3]); if $G \to S$ is an abelian scheme, one can take $V_{G/S}$ to be the dual of $F_{G^t/S}$, where G^t is the dual abelian scheme of G over S (this is immediate by the definition of G^t). Hence $V_{G/S}$ is a natural homomorphism when $G \to S$ is either a flat finite commutative group scheme or an abelian scheme, called the *Verschiebung* of G over S.

2.4. α-groups.

Let $\pi : G \to S$ be a flat finite commutative group scheme such that $(V_{G/S} : G^{(p)} \to G) = 0$. Let $\mathcal{F}$ be the subsheaf of $\pi_* O_G$ of sections s such that

$$m^*(s) = s \otimes 1 + 1 \otimes s.$$ (2.4.1)

Then canonically

$$\mathcal{F} \cong \mathrm{Lie}(G^D/S)$$ (2.4.2)

which is locally free. In this case $\mathcal{F}$ generates $\pi_* O_G$ as an O_S-algebra, and the induced map $\mathcal{F} \to \omega_{G/S}$ is surjective ([45, pp.338-339]). We will call $\mathcal{F}$ the α-sheaf of G, whose rank r is called the α-rank of G when S is connected (note that in this case the rank of G is p^r).

In particular, if $F_{G/S} = 0$ also, then G is called an α-group. In this case $\mathcal{F} \cong \omega_{G/S}$ canonically. It is well-known that any α-group of α-rank r is locally (i.e. over an open affine cover of S) isomorphic to

$$\overbrace{\alpha_p \times \ldots \times \alpha_p}^{r \text{ copies}} \times S$$ (2.4.3)

(cf. [45, p.339]).

Let $\mathcal{F}$ be a locally free sheaf of O_S-modules. Let $\mathcal{R} = \mathrm{Sym}(\mathcal{F})$, the symmetric algebra over O_S. We can define a group scheme structure on $G_{\mathcal{F}} = \mathbf{Spec}_S(\mathcal{R})$ by letting

$$m^*(s) = s \otimes 1 + 1 \otimes s, \quad \iota^*(s) = -s, \quad o^*(s) = 0$$ (2.4.4)

for any section s of $\mathcal{F}$. Then

$$G_{\mathcal{F}}^1 := \ker(F_{G_{\mathcal{F}}/S}) \cong \mathbf{Spec}_S(\mathcal{R}/\mathcal{F}^{[p]})$$ (2.4.5)

is an α-group, where $\mathcal{F}^{[p]}$ is the ideal sheaf of $\mathcal{R}$ generated by the pth powers of the sections of $\mathcal{F}$. We have ([45, p.339]):

17

Fact. *There is an anti-equivalence of categories*

$$((\text{flat coherent sheaves of } O_S\text{-modules})) \leftrightarrow ((\alpha\text{-groups over } S))$$
$$\text{the } \alpha\text{-sheaf of } G \;\leftarrowtail\; G$$
$$\mathcal{F} \;\mapsto\; G^1_{\mathcal{F}}$$

compatible with the Cartier dual functor. Hence any α-group is uniquely determined by its α-sheaf up to isomorphism.

2.5. The subgroup scheme $A(X)$ of a commutative group scheme X.

Let K be any field of characteristic p, not necessarily perfect.

Proposition. *For a commutative group scheme X of finite type over K, there is a smallest subgroup scheme $A(X) \subset X$ containing all of the images of $\alpha_p \to X$. Furthermore, for any field extension $L \supset K$ we have $A(X) \otimes_K L = A(X \otimes_K L) \subset X \otimes_K L$.*

Proof. Let $G = X[F]$. Let T be the set of sections s in the α-sheaf of G^D such that $s^p = 0$. Then $H = \operatorname{Spec} K[T]$ has a group scheme structure which is a quotient group scheme of G. Hence H^D can be identified with a closed subgroup scheme $A(X)$ of X. It is easy to see that $A(X) \otimes_K L = A(X \otimes_K L)$ for any field extension $L \supset K$.

If $f : \alpha_p \to X$ is a homomorphism. then f induces $f_0 : \alpha_p \to G$, hence induces a K-linear homomorphism ϕ from the α-sheaf of α_p^D to the α-sheaf of G^D. (The α-sheaf of α_p^D is isomorphic to K.) Clearly $\operatorname{im}(\phi) \subset T$, hence f_0^D factors through H, therefore f_0 factors through $A(X)$. On the other hand, every non-zero section of T gives a monomorphism $\alpha_p \hookrightarrow G \hookrightarrow X$, hence $A(X)$ is the smallest subgroup scheme of X containing all of the images of $\alpha_p \to X$. Q.E.D.

Note that $A(X)$ is an α-group over K. The α-rank of $A(X)$ is just $a(X)$ (see 1.5).

If K is perfect, then there is X_0 such that $X_0^{(p)} \cong X$. Clearly $A(X) = \ker(F_{X/K}) \cap \ker(V_{X_0/K})$. In this case one can define $a(X)$ by means of the Dieudonné module of X, see (5.2.7).

Remark. Over any field K, a *p-Lie algebra* is a Lie algebra $\mathfrak{g}$ together with a "pth power map" $x \mapsto x^p \in \mathfrak{g}$ such that $\operatorname{ad}(x^p) = (\operatorname{ad} x)^p$ and $(cx)^p = c^p x^p$ ($\forall c \in K$. $x \in \mathfrak{g}$). There is an equivalence of categories

$$((\text{finite group schemes } G \text{ over } K \text{ with } F_{G/K} = 0))/\text{isomorphisms} \xrightarrow{\lambda}$$
$$((p\text{-Lie algebras over } K))/\text{isomorphisms}$$

(cf. [7, II.7], [8, VII$_A$.7.4] or [55, p. 139]). For a commutative group scheme X of finite type over K, under this equivalence we have $\lambda(A(X)) = \{x \in \lambda(X[F]) | x^p = 0\}$. This is another way to give $A(X)$. In fact. $\lambda(X[F])$ is canonically isomorphic to the α-sheaf T of $X[F]^D$ with the trivial Lie algebra structure and the pth power map induced by the multiplication of $O_{X[F]^D}$.

18

3. Flag type quotients

3.1. An observation on minimal isogenies.

The definition of polarized flag type quotients is motivated by the following fact.

Let X be a *supergeneral* abelian variety of dimension g over k. Let $\rho : E^g \otimes k \to X$ be a minimal isogeny (see 1.8). Let $G = \ker(\rho)$ and

$$G_i = G \cap (E^g[F^{g-1-i}] \otimes k) \quad (0 \leq i \leq g - 1). \tag{3.1.1}$$

Then we get a sequence of isogenies

$$\rho_i : Y_i \to Y_{i-1} \quad (1 \leq i \leq g - 1), \tag{3.1.2}$$

where $Y_i = E^g \otimes k/G_i$. We have $Y_0 \cong X$ and $\ker(\rho_i)$ is an α-group of α-rank i $(1 \leq i \leq g - 1)$ (see [61, Theorem 2.2]). Furthermore, if X has a principal polarization, then it induces polarizations η_i on Y_i for each i such that

$$\ker(\eta_i) \subset \ker(F^{i-j} \circ V^j) \quad (\forall \, 0 \leq j \leq i/2), \tag{3.1.3}$$

where F, V are the relative Frobenius and Verschiebung respectively. In particular, $\ker(\eta_{g-1}) = Y_{g-1}[F^{g-1}]$ ([61, Theorem 3.1]).

3.2. Flag type quotient (FTQ).

Let S be an F_p-scheme.

Definition. A *flag type quotient (FTQ)* of dimension $g > 0$ over S consists of abelian schemes Y_i $(0 \leq i < g)$ of dimension g over S together with isogenies $\rho_i : Y_i \to Y_{i-1}$ $(0 < i < g)$ such that

i) $Y_{g-1} = E^g \times S$;
ii) $\ker(\rho_i)$ is an α-group (see 2.4) of α-rank i $(0 < i < g)$.

The FTQ $\{Y_{g-1} \to \ldots \to Y_0\}$ is called *rigid* if in addition that

iii) $\ker(Y_{g-1} \to Y_i) = \ker(Y_{g-1} \to Y_0) \cap Y_{g-1}[F^{g-1-i}]$ $(0 < i < g)$.

3.3. Examples of FTQs.

Example. i) Let $X = Y_0$ be a supergeneral abelian variety over k of dimension $g > 1$. Then up to isomorphism there exists a *unique* FTQ

$$Y_{g-1} \to \ldots \to Y_1 \to Y_0 = X \tag{3.3.1}$$

given by 3.1 above, and this is rigid ([61, Theorem 2.2]).

ii) Let $g = 3$. Choose any embedding $\alpha_p \times \alpha_p \hookrightarrow E^3 \otimes k$. Let

$$Y_1 := (E^3 \otimes k)/i(\alpha_p \times \alpha_p), \quad j : (E^3[F] \otimes k)/i(\alpha_p \times \alpha_p) \cong \alpha_p \hookrightarrow Y_1 \qquad (3.3.2)$$

and

$$Y_0 := Y_1/j(\alpha_p) \cong (E^3/E^3[F]) \otimes k. \qquad (3.3.3)$$

Then

$$Y_2 = E^3 \otimes k \to Y_1 \to Y_0 \qquad (3.3.4)$$

is an FTQ which is not rigid. On the other hand, for any embedding $j' : \alpha_p \hookrightarrow Y_1$ such that $j'(\alpha_p) \neq j(\alpha_p)$, the FTQ

$$Y_2 = E^3 \otimes k \to Y_1 \to Y_0' := Y_1/j'(\alpha_p) \qquad (3.3.5)$$

is rigid. (For further details on $g = 3$, see Remark 6.4 and Example 9.4.)

3.4. For the terminology "rigid FTQ".

Remark. We use the word "rigid" because of the following reason. If $Y_0 = Y \times S$ for some abelian variety Y over K and some connected K-scheme S, then the FTQ is constant if it is rigid. This is clear because a rigid FTQ is uniquely determined by $Y_{g-1} \to Y_0$, and $Y_{g-1} \to Y_0$ is induced by some homomorphism $E^g \otimes K \to Y$ by the Rigidity Lemma.

3.5. Universal FTQ.

Lemma. *The functor*

$$\mathfrak{q}_g \colon ((\mathbb{F}_p\text{-schemes})) \longrightarrow ((\text{sets}))$$
$$S \mapsto \{ \text{FTQs of dimension } g \text{ over } S \} \,/\text{isomorphisms}$$

is represented by a projective scheme $\mathcal{Q}_g$ over $\mathbb{F}_p$. Also the functor

$$\mathfrak{q}_g' \colon ((\mathbb{F}_p\text{-schemes})) \longrightarrow ((\text{sets}))$$
$$S \mapsto \{ \text{rigid FTQs of dimension } g \text{ over } S \} \,/\text{isomorphisms}$$

is represented by a quasi-projective scheme $\mathcal{Q}_g'$ over $\mathbb{F}_p$. By the canonical morphism $\mathcal{Q}_g' \to \mathcal{Q}_g$, we can identify $\mathcal{Q}_g'$ with an open subscheme of $\mathcal{Q}_g$.

Proof. We set up the fine moduli scheme inductively. Let $\mathfrak{U}_m$ be the category of "m-truncated FTQs", i.e. sequences of isogenies $\{\rho_i : Y_i \to Y_{i-1} \ (m < i < g)\}$ satisfying i) and ii) (for $i > m$ only) in the definition of FTQs. We will show that each $\mathfrak{U}_m$ has a fine moduli scheme $\mathcal{U}_m$ over $\mathbb{F}_p$. Then clearly $\mathcal{Q}_g = \mathcal{U}_0$ is the fine moduli scheme of FTQs.

Obviously $\mathfrak{U}_{g-1}$ is represented by $\mathrm{Spec}(\mathbb{F}_p)$. We also note that given an m-truncated FTQ $\{\rho_i : Y_i \to Y_{i-1} \ (m < i < g)\}$ over some S, to extend it to an $(m-1)$-truncated FTQ is equivalent to giving a closed subgroup scheme of Y_m over S which is an α-group of α-rank m.

Suppose we have shown the existence of $\mathcal{U}_m$. Let $\{Y_i(m \leq i < g); \rho_i(m < i < g)\}$ be the universal m-truncated FTQ over $\mathcal{U}_m$. Then for any $(m-1)$-truncated

FTQ over any base S, its m-truncation induces a morphism $S \to \mathcal{U}_m$. Let $G = \ker(F_{Y_m/\mathcal{U}_m})$. By Fact 2.2, the functor

$$((\mathcal{U}_m\text{-schemes})) \longrightarrow ((\text{sets}))$$
$$S \mapsto \{\text{closed subgroup schemes of } G \times_{\mathcal{U}_m} S \text{ over } S$$
$$\text{which is an } \alpha\text{-group of } \alpha\text{-rank } m\}$$

is represented by a relatively projective scheme $\mathcal{U}_{m-1}$ over $\mathcal{U}_m$. (We leave to the reader to check that it is an algebraic condition for a subgroup scheme of $G \times_{\mathcal{U}_m} S$ to be an α-group.) By the above note, we see that $\mathcal{U}_{m-1}$ is just the fine moduli scheme of $(m-1)$-truncated FTQs.

The remaining statements are clear if one notices that condition iii) in Definition 3.2 is an open condition. Q.E.D.

3.6. Polarized flag type quotient (PFTQ).

We note that by (1.2.4), every polarization of $E^g \otimes k$ is defined over $\mathbf{F}_{p^2}$ (cf. [56, Definition 6.3]).

Let η be a polarization of $E^g \otimes K$ over K such that

$$\ker(\eta) = E^g[F^{g-1}] \otimes K. \tag{3.6.1}$$

Such an isogeny exists for any $K \supset \mathbf{F}_{p^2}$ (it even exists for $K = \mathbf{F}_p$ when g is odd). Let S be a K-scheme.

Definition. A *polarized flag type quotient (PFTQ)* over S with respect to η consists of polarized abelian schemes $\{Y_i, \eta_i\}$ $(0 \le i < g)$ of dimension g over S together with isogenies $\rho_i : Y_i \to Y_{i-1}$ $(0 < i < g)$ compatible with the polarizations (i.e. $\rho_i^t \circ \eta_{i-1} \circ \rho_i = \eta_i$ for all $0 < i < g$) such that $\{Y_i \ (0 \le i < g); \rho_i \ (0 < i < g)\}$ is an FTQ and

a) $\eta_{g-1} = \eta \times \mathrm{id}_S$;
b) $\ker(\eta_i) \subset \ker(F^{i-j} \circ V^j)$ $(\forall 0 \le j \le [i/2])$, where F, V are the relative Frobenius and Verschiebung respectively.

(In particular, η_0 is a principal polarization.) The PFTQ is called *rigid* if the corresponding FTQ is so.

3.7. Universal PFTQ.

Let $X \to S$ be an abelian scheme, and $\eta : X \to X^t$ be a polarization. By the duality theorem (cf. [64, Theorem 18.1] or [55, p.143]) we have a commutative diagram

$$
\begin{array}{ccccccccc}
0 \to & G := \ker(\eta) & \longrightarrow & X & \xrightarrow{\eta} & X^t & \to 0 \\
& \downarrow{\scriptstyle \cong} & & \downarrow{\scriptstyle \cong} & & \downarrow{\scriptstyle \mathrm{id}} & & \\
0 \to & G^D \cong \ker(\eta^t) & \longrightarrow & X^{tt} & \xrightarrow{\eta^t} & X^t & \to 0
\end{array}
\tag{3.7.1}
$$

and we see that η induces a canonical isomorphism $\theta : G \to G^D$ such that $\theta^D \circ \theta = \mathrm{id}_G : G \to G^{DD} \cong G$.

Lemma. *Let η be a polarization of $E^g \odot K$ over K satisfying (3.6.1). Then the functor*

$$\mathfrak{p}_\eta : ((K\text{-schemes}))\longrightarrow((\text{sets}))$$
$$S \mapsto \{\text{ PFTQs over } S \text{ with respect to } \eta\} /\text{isomorphisms}$$

is represented by a projective scheme $\mathcal{P}_{g,\eta}$ over K. Also the functor

$$\mathfrak{p}'_\eta : ((K\text{-schemes}))\longrightarrow((\text{sets}))$$
$$S \mapsto \{\text{ rigid PFTQs over } S \text{ with respect to } \eta\} /\text{isomorphisms}$$

is represented by a quasi-projective scheme $\mathcal{P}'_{g,\eta}$ over K. By the canonical morphism $\mathcal{P}'_{g,\eta} \to \mathcal{P}_{g,\eta}$, we can identify $\mathcal{P}'_{g,\eta}$ with an open subscheme of $\mathcal{P}_{g,\eta}$.

Furthermore, the FTQ structure of the universal PFTQ over $\mathcal{P}_{g,\eta}$ induces a closed immersion $\mathcal{P}_{g,\eta} \hookrightarrow \mathcal{Q}_g \odot K$. If we identify $\mathcal{P}_{g,\eta}$ with a closed subscheme of $\mathcal{Q}_g \odot K$ by this, then we have

$$\mathcal{P}'_{g,\eta} = \mathcal{P}_{g,\eta} \cap (\mathcal{Q}'_g \odot K). \tag{3.7.2}$$

Proof. We can make use of the proof of Lemma 3.5, with additional consideration of polarizations.

Suppose we are given an m-truncated PFTQ $\{Y_i, \eta_i(m \le i < g); \rho_i(m < i < g)\}$ over some S. Let $G = \ker(\eta_m)$. Then η_m induces an isomorphism $\theta : G \to G^D$ as above. Let $f : H \hookrightarrow G$ be a closed subgroup scheme which is an α-group of α-rank m and $Y_{m-1} = Y_m/H$. We claim that η_m induces a polarization on Y_{m-1} iff

$$f^D \circ \theta \circ f = 0. \tag{3.7.3}$$

(In other words, H is "isotropic" with respect to θ.) Indeed, (3.7.3) is obviously necessary. Conversely, if (3.7.3) holds, then η_m induces a homomorphism $\eta_{m-1} : Y_{m-1} \to Y^t_{m-1}$, which is a polarization by descent theory (see [55, p.231]).

To show the existence of the fine moduli scheme of $(m-1)$-truncated PFTQs with induction hypothesis, it is enough to check that (3.7.3) and condition b) in the definition of PFTQs are both closed conditions. This is again left to the reader.

The remaining statements are clear. Q.E.D.

We will denote by $\mathfrak{P}_{g,\eta}$ (resp. $\mathfrak{P}'_{g,\eta}$) the category of PFTQs (resp. rigid PFTQs) of dimension g with respect to η, and denote by $\mathfrak{Q}_g$ (resp. $\mathfrak{Q}'_g$) the category of FTQs (resp. rigid FTQs) of dimension g.

3.8. Examples of moduli of FTQs and of PFTQs.

Example. Clearly $\mathcal{P}_{1,\eta} \cong \mathrm{Spec}(K)$ and $\mathcal{Q}_1 \cong \mathrm{Spec}(\mathbb{F}_p)$. For $g = 2$, note that to give a PFTQ with respect to η over S is equivalent to giving an α-subgroup $H \subset \ker(\eta) \times S$ of α-rank 1, because H is automatically isotropic. Furthermore, to give such an H is equivalent to giving a flat quotient of the α-sheaf of $\ker(\eta) \times S$ ($\cong O_S^{\oplus 2}$) of rank 1 over O_S. Hence $\mathcal{P}_{2,\eta} \cong \mathbb{P}^1_K$. Similarly, $\mathcal{Q}_2 \cong \mathbb{P}^1_{\mathbb{F}_p}$.

3.9. IFTQ of group schemes.

The following definition simplifies our study on PFTQs.

Let η be a polarization of $E^g \otimes \mathsf{F}_{p^2}$ satisfying (3.6.1) for $K = \mathsf{F}_{p^2}$ (i.e. $\ker(\eta) = E^g[F^{g-1}] \otimes \mathsf{F}_{p^2}$) and $G = E^g[F^{g-1}] \otimes \mathsf{F}_{p^2}$. Then η induces an isomorphism $\theta : G \to G^D$ as in 3.7. We have an immediate consequence of Corollary 6.1 (see (1.2.4) also):

Lemma. *Up to equivalence θ is independent of the choice of η, i.e. if η' is another polarization of $E^g \otimes \mathsf{F}_{p^2}$ satisfying $\ker(\eta') = G$ and $\theta' : G \to G^D$ is the isomorphism induced by η', then there is an automorphism ϕ of G such that $\phi^D \circ \theta \circ \phi = \theta'$.*

We fix an η (and θ) over F_{p^2}.

Definition. *An isotropic flag type quotient (IFTQ) of finite group schemes of genus g (with respect to θ) over an F_{p^2}-scheme S is a filtration $0 = G_{g-1} \subset ... \subset G_0 \subset G \times S$ of flat subgroup schemes of $G \times S$ over S such that*

    i) G_{i-1}/G_i is an α-group of α-rank i $(0 < i < g)$;
    ii) $f_i^D \circ \theta_S \circ f_i = 0$, where $f_i : G_i \to G$ is the inclusion, and $F^{i-j} \circ V^j = 0$ on $\ker(f_i^D \circ \theta_S)/G_i$ $(0 \le j \le i/2)$, where $\theta_S = \theta \times \mathrm{id}_S$;

which is called rigid if in addition that

    iii) $G_i = G_0 \cap (G[F^{g-1-i}] \times S)$ $(0 < i < g)$.

3.10. Universal IFTQ.

Let $\mathfrak{P}_g$ be the category of IFTQs of group schemes of genus g, and $\mathfrak{P}'_g$ be the category of rigid IFTQs of group schemes of genus g. Note that the definition of PFTQs (Definition 3.6) only concerns subgroup schemes of G. We see that $\mathfrak{P}_g$ is equivalent to $\mathfrak{P}_{g,\eta}$. Hence we have

Corollary. *The category $\mathfrak{P}_g$ has a fine moduli scheme $\mathcal{P}_g$, and $\mathfrak{P}'_g$ has a fine moduli scheme $\mathcal{P}'_g$ which can be identified with an open subscheme of $\mathcal{P}_g$.*

Note that up to isomorphism $\mathcal{P}_g$ is independent of the choice of θ, and $\mathcal{P}_{g,\eta'} \cong \mathcal{P}_g$ for any η' satisfying $\ker(\eta') = E^g[F^{g-1}] \otimes \mathsf{F}_{p^2}$, by the Lemma 3.9.

Let $\{G_{g-1} \subset ... \subset G_0\}$ be an IFTQ of group schemes over any $K \supset \mathsf{F}_{p^2}$ and $\{Y_i, \eta_i(0 \le i < g); \rho_i(0 < i < g)\}$ be the corresponding PFTQ with respect to η. Then $Y_i \cong E^g \otimes K/G_i$ $(0 \le i \le g-1)$. Therefore $a(Y_0) = 1$ iff $a(G_0) = 1$ (see Fact 5.6).

4. Main results on $\mathcal{S}_{g,1}$
(the principally polarized case)

In this chapter we collect the main propositions about the supersingular locus $\mathcal{S}_{g,1} \subset \mathcal{A}_{g,1} \otimes \mathbf{F}_p$ which will be proved separately in Chapter 6, Chapter 7 and in Chapter 8, and we prove the main theorem on the structure of $\mathcal{S}_{g,1}$ using these propositions. We also explain some strategical techniques which we will use in the proofs of the propositions.

4.1. Existence of rigid PFTQ for a principally polarized supersingular abelian variety.

Proposition. *Let (X, λ) be a principally polarized supersingular abelian variety of dimension g over k. Then there is a rigid PFTQ over k (with respect to some η over k) $\{Y_i \ (0 \le i \le g-1); \rho_i \ (1 \le i \le g-1)\}$ such that $(Y_0, \eta_0) \cong (X, \lambda)$. Furthermore, there are only a finite number of such PFTQs up to isomorphism.*

This is an immediate consequence of Proposition 6.3.

4.2. A parameter space of principally polarized supersingular abelian varieties.

Two polarizations μ and μ' of an abelian variety X are called *equivalent* if there is an automorphism ϕ of X such that $\phi^t \circ \mu \circ \phi = \mu'$.

Let $K = \bar{\mathbf{F}}_p$. Let Λ be a set of representatives of equivalence classes of polarizations η of $E^g \otimes K$ satisfying (3.6.1). Then we have a canonical morphism

$$\Psi : \coprod_{\eta \in \Lambda} \mathcal{P}'_{g,\eta} \to \mathcal{S}_{g,1} \otimes \bar{\mathbf{F}}_p, \tag{4.2.1}$$

where $\mathcal{S}_{g,1}$ is the supersingular locus in $\mathcal{A}_{g,1}$ (see 1.10). Note that Ψ is defined over $\mathbf{F}_{p^2}$ by 3.6 and Lemma 3.7. From Proposition 4.1 we get

Corollary. *The morphism Ψ is quasi-finite and surjective.*

4.3. The important properties of $\mathcal{P}'_g$.

Proposition. *Let $\{G_{g-1} \subset ... \subset G_0\}$ be the universal IFTQ of group schemes over $\mathcal{P}_g$ (see Corollary 3.10).*

    i) $\mathcal{P}'_g$ is non-singular and geometrically integral of dimension $[g^2/4]$.

    ii) The generic fiber of G_0 over $\mathcal{P}'_g$ has a-number equal to 1.

In Chapter 7 we give a proof of the "*Weak Form of Proposition 4.3*", i.e. replacing i) by i') $\mathcal{P}'_g$ is geometrically irreducible of dimension $[g^2/4]$.

And we sketch a proof of the fact that $\mathcal{P}'_g$ is non-singular in 11.3; We also give a complete proof of this fact for $g = 4$ in 9.7.

4.4. The important properties of $\mathcal{P}'_{g,\eta}$.

Corollary. Let η be a polarization of $E^g \otimes K$ satisfying (3.6.1). Let $\{Y_i \ (0 \leq i < g); \rho_i \ (0 < i < g)\}$ be the universal PFTQ over $\mathcal{P}'_{g,\eta}$ (see Lemma 3.7).

  i) $\mathcal{P}'_{g,\eta}$ is non-singular and geometrically integral of dimension $[g^2/4]$.
  ii) The generic fiber of Y_0 over $\mathcal{P}'_{g,\eta}$ is supergeneral.

4.5. About the structure of $\mathcal{P}_g$.

Remark. Note that $\mathcal{P}_g$ is integral and non-singular for $g \leq 3$ (see Examples 3.8 and Example 9.4). Note also that the subscheme $\mathcal{P}'_g$ is integral and non-singular for any g. However, in general $\mathcal{P}_g$ is neither non-singular nor irreducible (see Example 9.6).

4.6. Some class numbers, $H_g(p,1)$.

Let B be the definite quaternion algebra over $\mathbf{Q}$ with discriminant p (as in (1.2.5)). Let $\mathcal{O}$ be a maximal order of B (as in (1.2.4)). By a theorem of Eichler (cf. [92, Lemma 4.4]), every left $\mathcal{O}$-lattice in $B^{\oplus g}$ is equal to $\mathcal{O}^{\oplus g}x$ for some $x \in GL_g(B)$. Let

$$G = \{h \in M_g(B) | h\bar{h}^t = rI \text{ for some } r \in \mathbf{Q}^\times\}. \tag{4.6.1}$$

Two $\mathcal{O}$-lattices L and L' in $B^{\oplus g}$ are called *globally equivalent* (denoted by $L \sim L'$) if there exists $h \in G$ such that $L' = Lh$.

  Let

$$\Sigma := \{f \in M_g(\mathcal{O}) | f = (\bar{f})^t \text{ is positive definite}\}. \tag{4.6.2}$$

Two elements $f, f' \in \Sigma$ are called *quasi-equivalent* (denoted by $f \sim f'$) if there exists $\gamma \in GL_g(\mathcal{O})$ and a positive rational number m such that $\bar{\gamma}^t f \gamma = m f'$, and f and f' are called *equivalent* (denoted by $f \approx f'$) if in addition we have $m = 1$.

  By the argument of [31, Lemma 2.5], the map $x \mapsto x\bar{x}^t$ induces a one-to-one correspondence between the global equivalence classes of left $\mathcal{O}$-lattices in $B^{\oplus g}$ and $\Sigma/\sim$.

  Let

$$N_p = \mathcal{O}_p^{\oplus g} \begin{pmatrix} 1_{g-r} & 0 \\ 0 & \pi 1_r \end{pmatrix} \xi, \tag{4.6.3}$$

where π is a prime element of $\mathcal{O}_p$ and $\xi \in GL_g(B_p)$ such that

$$\xi\bar{\xi}^t = \begin{pmatrix} \mathbf{0} & & 1 \\ & \cdot & \\ & \cdot & \\ 1 & & \mathbf{0} \end{pmatrix}. \tag{4.6.4}$$

25

Let $\mathcal{L}_g(p,1)$ be the set of left $\mathcal{O}$-lattices L in $B^{\oplus g}$ such that $L_l \sim \mathcal{O}_l^{\oplus g}$ for any prime number l, and $\mathcal{L}_g(1,p)$ be the set of left $\mathcal{O}$-lattices L in $B^{\oplus g}$ such that $L_p \sim N_p$ and $L_l \sim \mathcal{O}_l^{\oplus g}$ for any prime number $l \neq p$. By definition, the class number $H_g(p,1)$ (resp., $H_g(1,p)$) of the principal genus (resp. the non-principal genus) of the hermitian space $B^{\oplus g}$ is the number of global equivalence classes in $\mathcal{L}_g(p,1)$ (resp., $\mathcal{L}_g(1,p)$) (cf. [25, p.696]). We quote [31, Theorem 2.10]:

Fact. *The number of equivalence classes of principal polarizations of $E^g \otimes \bar{\mathsf{F}}_p$ is equal to $H_g(p,1)$.*

4.7. The class number $H_g(1,p)$.

We also have (see 8.8 and 8.9):

Proposition. *The class number $H_g(1,p)$ is equal to the number of equivalence classes of polarizations η of $E^g \otimes \bar{\mathsf{F}}_p$ such that $\ker(F) \subset \ker(\eta)$ and $\deg(\eta) = p^{2r}$, where $r = [(g+1)/2]$.*

4.8. Polarizations and class numbers.

Corollary. *Let $n \in \mathbb{Z}_{>0}$ and $r = [(g+1)/2]$.*

    i) The number of equivalence classes of polarizations η of $E^g \otimes \bar{\mathsf{F}}_p$ such that $\ker(\eta) = \ker(p^n \cdot)$ is equal to $H_g(p,1)$.

    ii) The number of equivalence classes of polarizations η of $E^g \otimes \bar{\mathsf{F}}_p$ such that $\ker(F^{2n+1}) \subset \ker(\eta)$ and $\deg(\eta) = p^{2(ng+r)}$ is equal to $H_g(1,p)$.

    iii) The number of equivalence classes of polarizations η of $E^g \otimes \bar{\mathsf{F}}_p$ such that $\ker(\eta) \subset \ker(F^{2n+1})$ and $\deg(\eta) = p^{2(ng+g-r)}$ is also equal to $H_g(1,p)$.

4.9. The main theorem on $\mathcal{S}_{g,1}$.

Summarizing 4.1-4.8 we get

Theorem. *For any $g > 0$ and any prime number p we have:*

    i) Every irreducible component of $\mathcal{S}_{g,1}$ has dimension $[g^2/4]$.

    ii) $\#\{\text{irreducible components of } \mathcal{S}_{g,1} \otimes \bar{\mathsf{F}}_p\} = \begin{cases} H_g(p,1) & \text{if } g \text{ is odd} \\ H_g(1,p) & \text{if } g \text{ is even} \end{cases}$.

    iii) $a(-/V) = 1$ (see 1.10) for each irreducible component V of $\mathcal{S}_{g,1}$.

Proof. Since Ψ is surjective by Corollary 4.2, and each $\mathcal{P}'_{g,\eta}$ is irreducible by Corollary 4.4.i), we see that $\mathcal{S}_{g,1} \otimes \bar{\mathsf{F}}_p$ is the union of the closures of $\Psi(\mathcal{P}'_{g,\eta})$'s, which are irreducible components. Since Ψ is quasi-finite by Corollary 4.2, we see that

$$\dim(\Psi(\mathcal{P}'_{g,\eta})) = \dim(\mathcal{P}'_{g,\eta}) = [\frac{g^2}{4}] \qquad (4.9.1)$$

by Corollary 4.4.i). Thus i) is proved.

26

We note that for any PFTQ $\{Y_i\,(0 \le i < g); \rho_i\,(0 < i < g)\}$ over $\bar{\mathsf{F}}_p$ with respect to some η, if $a(Y_0) = 1$, then η is uniquely determined by η_0 up to equivalence, because $Y_{g-1} \to Y_0$ is minimal (see 1.8 and 3.1). By Corollary 4.4.ii), we see that $\Psi(\mathcal{P}'_{g,\eta})$ has a geometric point which represents a supergeneral abelian variety, hence η is uniquely determined by $\Psi(\mathcal{P}'_{g,\eta})$ up to equivalence. Therefore Ψ induces a one to one correspondence between Λ and the set of irreducible components of $\mathcal{S}_{g,1} \otimes \bar{\mathsf{F}}_p$. This shows

$$\#\{\text{irreducible components of } \mathcal{S}_{g,1} \otimes \bar{\mathsf{F}}_p\} = \#(\Lambda). \tag{4.9.2}$$

Hence ii) is proved by applying Corollary 4.8.i) when g is odd and Corollary 4.8.ii) when g is even.

Finally, iii) is immediate by Corollary 4.4.ii).   Q.E.D.

4.10. A birational morphism.

Remark. For any $K \supset \mathsf{F}_{p^2}$ and any η satisfying (3.6.1), let $\mathrm{Aut}(E^g \otimes K, \eta)$ be the finite group of automorphisms ϕ of $E^g \otimes K$ such that $\phi^t \circ \eta \circ \phi = \eta$. There is an obvious action of $\mathrm{Aut}(E^g \otimes K, \eta)$ on $\mathcal{P}_{g,\eta}$. The morphism $\mathcal{P}'_{g,\eta} \to \mathcal{S}_{g,1}$ factors through $\mathcal{P}'_{g,\eta}/\mathrm{Aut}(E^g \otimes K, \eta)$. This gives a birational equivalence from $\mathcal{P}'_{g,\eta}/\mathrm{Aut}(E^g \otimes K, \eta)$ to an irreducible component of $\mathcal{S}_{g,1} \otimes K$.

In general cases the group $\mathrm{Aut}(E^g \otimes K, \eta)$ depend on η. These groups were completely classified for $g = 2$ by Ibukiyama ([32, Theorem 7.1]).

4.11. Some strategical techniques for the proofs of the propositions.

i) The fact that $\mathcal{P}'_g$ is non-singular in Proposition 4.3 is not so easy to show. However, in the proof of Theorem 4.9 we only need to use the (geometric) irreducibility of $\mathcal{P}'_g$. We will only give a complete proof of the Weak Form of Proposition 4.3 (see 4.3 above), and sketch a proof of Proposition 4.3 after the proof of Proposition 10.1. We also give a complete proof of Proposition 4.3 for $g = 4$ (see 9.7), which can help the reader to work out a complete proof (for general g) from the sketch.

ii) For the proof of of the Weak Form of Proposition 4.3, we can use an argument on Dieudonné modules which is much easier and more intuitive.

We can also use Dieudonné module arguments to prove Proposition 4.1, because it only concerns supersingular abelian varieties on which the Dieudonné module functor D (see 5.2) works well.

iii) In the proof of Proposition 4.3 we will use double induction, namely we first take an open cover $\{\mathcal{U}^\Theta\}$ of $\mathcal{P}'_g$, then for each Θ we get a morphism $\mathcal{U}^\Theta \to \mathcal{P}'_{g-2}$. In this way we can apply induction from g to $g + 2$. We then factor $\mathcal{U}^\Theta \to \mathcal{P}'_{g-2}$ to a sequence of morphisms

$$\mathcal{U}^\Theta = \mathcal{U}_0 \to \ldots \to \mathcal{U}_{g-1} = \mathcal{P}'_{g-2}. \tag{4.11.1}$$

Then the key step is to show that each morphism $\mathcal{U}_{m-1} \to \mathcal{U}_m$ in (4.11.1) is smooth of relative dimension 1.

5. Prerequisites about Dieudonné modules

In this chapter we collect and briefly explain the material about Dieudonné modules which we will use. These can be found in the literature, and there are several different (but equivalent) formulations. However, some of the theories give covariant functors (from group schemes to modules, see e.g. [43, p. 91]) while the others give contravariant functors (see e.g. [6, pp. 63-71]). In this book we use *contravariant* theory. For those readers who are accustomed with covariant theory, it is not hard to convert our argument using duality of commutative formal groups.

5.1. The Witt scheme.

Let K be a perfect field. Let $W = W(K)$ be the ring of (infinite) Witt vectors of K (cf. [85 §II.6]). Up to isomorphism W is the unique complete DVR with maximal ideal (p) such that $W/pW \cong K$. Furthermore, the absolute Frobenius of K can be uniquely lifted to an automorphism σ of W.

There is a way to express Witt vectors as sequences in K such that for any $u = (u_0, u_1, ...), v = (v_0, v_1, ...) \in W$ $(u_i, v_i \in K \ \forall i)$,

$$u + v = (\phi_0, \phi_1, ...), \quad uv = (\psi_0, \psi_1, ...) \tag{5.1.1}$$

where ϕ_n and ψ_n are polynomials in $u_0, ... u_n, v_0, ..., v_n$ for each n. Furthermore,

$$u^\sigma = (u_0^p, u_1^p, ...), \quad pu = (0, u_0^p, u_1^p, ...). \tag{5.1.2}$$

One can use (5.1.1) to define an addition and a product on $\mathbb{A}_K^n$ for each n. In this way $\mathbb{A}_K^n$ is endowed a ring scheme structure, denoted by $\mathcal{W}_n$ (cf. [54, Lecture 26]). Define the formal limit

$$\mathcal{W} := \varinjlim_n \mathcal{W}_n \tag{5.1.3}$$

which is a formal ring scheme, called the *Witt scheme* over K. We denote by $\mu : \mathcal{W} \times \mathcal{W} \to \mathcal{W}$ the multiplication of $\mathcal{W}$.

Note that $W = W(K) \cong \mathcal{W}(K)$ (the set of K-points of $\mathcal{W}$) as a ring.

5.2. Dieudonné modules over a perfect field K.

For any $G \in \mathrm{Ob}(\mathfrak{C}_K)$ (where $\mathfrak{C}_K$ is the category of commutative formal groups over K, see 1.3) such that both G and G^t are infinitesimal, let

$$D(G) = \mathrm{Hom}_K(G, \mathcal{W}). \tag{5.2.1}$$

Then any $m \in D(G)$ and $a \in W$ (viewed as a K-point $\mathrm{Spec}(K) \to \mathcal{W}$) has a "product"

$$am : G \cong \mathrm{Spec}(K) \times G \xrightarrow{(a,m)} \mathcal{W} \times \mathcal{W} \xrightarrow{\mu} \mathcal{W} \tag{5.2.2}$$

28

in $D(G)$. Thus μ induces a W-module structure on $D(G)$.

The Frobenius and Verschiebung of G induce W-semilinear endomorphisms F and V of $D(G)$ respectively such that

$$F(ax) = a^\sigma F(x), \quad V(ax) = a^{\sigma^{-1}} V(x) \quad (\forall a \in W, x \in D(G)) \tag{5.2.3}$$

and

$$F \circ V = V \circ F = p \cdot \mathrm{id}_{D(G)}. \tag{5.2.4}$$

(Note that $F_{G/K} : G \to G^{(p)}$ induces a W-linear map $D(G)^{(p)} \cong D(G^{(p)}) \to D(G)$, which is equivalent to a W-semilinear map $F : D(G) \to D(G)$ as above.)

The formulation above can be generalized to any object $G \in \mathrm{Ob}(\mathfrak{C}_K)$ (see e.g. [6, pp. 63-71]), but we will only use the above case.

Let $A_K = W[F, V] = W(K)[F, V]$ be the associative ring satisfying the following defining relations:

a) $FV = VF = p$;

b) $Fa = a^\sigma F, \ Va = a^{\sigma^{-1}} V \ (\forall a \in W)$

and $A = \varprojlim_n A_K / p^n A_K$.

Definition. A *Dieudonné module* over W is a left A-module which is finitely generated as a W-module.

The above D can be viewed as a contravariant functor from $\mathfrak{C}_K$ to the category of Dieudonné modules over W. The following facts are well known (see e.g. [6, pp. 63-71] and [48. p.37).

Fact. *For any perfect field K we have:*

i) *The functor D is an anti-equivalence of categories from $\mathfrak{C}_K$ to the category of Dieudonné modules over $W = W(K)$.*

ii) *For any $G \in \mathrm{Ob}(\mathfrak{C}_K)$, there is a canonical isomorphism*

$$D(G^t) \cong \mathrm{Hom}_W(D(G), W) \tag{5.2.5}$$

where the A-module structure of $\mathrm{Hom}_W(D(G), W)$ is given by

$$(Ff)(x) = f(Vx)^\sigma, \quad (Vf)(x) = f(Fx)^{\sigma^{-1}} \tag{5.2.6}$$
$$(\forall x \in D(G), \ f \in \mathrm{Hom}_W(D(G), W)).$$

iii) *If $K = k$ is an algebraically closed field, then the p-divisible group $G_{m,n}$ in 1.4 satisfies*

$$D(G_{m,n}) \cong A/A(F^n - V^m), \tag{5.2.7}$$

where A and the left ideal $A(F^n - V^m)$ are viewed as left A-modules.

iv) *If X is an abelian variety of dimension g over K, then the Dieudonné module $D(\varphi_p X)$ of the Barsotti-Tate group $\varphi_p X$ (see (1.4.1)) is free of rank $2g$ over W.*

By slight abuse of notation we will denote $D(\varphi_p X)$ simply by $D(X)$, and call it the Dieudonné module of X. We say $D(X)$ has genus g if $\dim X = g$, i.e. $\text{rank}_W D(X) = 2g$.

The a-number of a Dieudonné module M is defined by

$$a(M) = \dim_K(M/(F, V)M). \tag{5.2.8}$$

By 2.5, for any commutative group scheme X of finite type over K, we have $a(D(X)) = a(X)$.

5.3. Dieudonné crystals.

The formulation above can also be generalized over a base S instead of $\text{Spec}(K)$. For example, we can use the theory of crystalline cohomology (cf. [1, Section 3] or [50, IV]).

Let S be a scheme of finite type over a perfect field K and $W = W(K)$. Denote by $(S/W)_{cris}$ the "site" (which is a category) of "S-PD thickenings" $\{U \hookrightarrow T, \delta\}$, where

a) U is an S-scheme;

b) $U \hookrightarrow T$ is a closed immersion over W with ideal sheaf I such that $p^n I = 0$ for some n;

c) δ is a collection of maps $\delta_i : I \to O_T$ ($i \in \mathbb{Z}_{\geq 0}$) such that for any sections x, y of I and λ of O_T (on one and the same open subset of T)

c1) $\delta_0(x) = 1$, $\delta_1(x) = x$, and $\delta_n(x)$ is in I when $n > 1$;

c2) $\delta_n(x + y) = \sum_{i+j=n} \delta_i(x)\delta_j(y)$;

c3) $\delta_n(\lambda x) = \lambda^n \delta_n(x)$;

c4) $\delta_i(x)\delta_j(x) = \binom{i+j}{j}\delta_{i+j}(x)$;

c5) $\delta_i(\delta_j(x)) = \frac{(ij)!}{i!(j!)^p}\delta_{ij}(x)$.

(Note that $n!\delta_n(x) = x^n$.) For example, if $U = \text{Spec}(R)$, then $\text{Spec} W_n(R)$ is an S-PD thickening of U for any $n > 0$, where $W_n(R)$ is the truncated Witt algebra of R.

Definition. A crystal on $(S/W)_{cris}$ is a sheaf of $O_{S/W}$-modules $\mathcal{E}$ such that for any morphism $f : \{U, T, \delta\} \to \{U', T', \delta'\}$ in $(S/W)_{cris}$, the canonical morphism $f^*\mathcal{E}_{\{U',T',\delta'\}} \to \mathcal{E}_{\{U,T,\delta\}}$ is an isomorphism.

For example, if $\mathcal{F}$ is a coherent sheaf on S, then we can get a crystal $\Phi^*\mathcal{F}$ as follows. For any object $\{U, T, \delta\}$ in $(S/W)_{cris}$, there is a unique morphism $\Phi_T : \text{Spec}_U(O_T/pO_T) \to U$ such that the composition of the inclusion

$$U \hookrightarrow \text{Spec}_U(O_T/pO_T)$$

with Φ_T is equal to the absolute Frobenius of $\text{Spec}_U(O_T/pO_T)$. Define

$$(\Phi^*\mathcal{F})_{\{U,T,\delta\}} = \Phi_T^*\pi^*\mathcal{F} \tag{5.3.1}$$

where $\pi : U \to S$ is the projection. Clearly we have $p\Phi^*\mathcal{F} = 0$.

A morphism of K-schemes $f : S' \to S$ induces a morphism $(S'/W)_{cris} \to (S/W)_{cris}$, which gives every crystal $\mathcal{E}$ on $(S/W)_{cris}$ a pull-back $f^*\mathcal{E}$ on $(S/W)_{cris}$. In particular, the pull-back of $\mathcal{E}$ via the absolute Frobenius σ of S will be denoted by $\mathcal{E}^{(p)}$.

Definition. A *Dieudonné crystal* on $(S/W)_{cris}$ is a coherent crystal $\mathcal{E}$ together with two $O_{S/W}$-linear morphisms $F : \mathcal{E}^{(p)} \to \mathcal{E}$ and $V : \mathcal{E} \to \mathcal{E}^{(p)}$ (called the relative Frobenius and Verschiebung respectively) such that

$$F \circ V = p \cdot \mathrm{id}_{\mathcal{E}}, \quad V \circ F = p \cdot \mathrm{id}_{\mathcal{E}^{(p)}}. \tag{5.3.2}$$

5.4. The Dieudonné crystal of a group scheme.

The following generalizes some results of Dieudonné modules over W (see [1, Section 3.3, 4.3, 5.3]).

Fact. *There is a right exact faithful contravariant functor* $\mathbf{D}$ *from* $\mathfrak{C}_S$ *(the category of commutative formal groups over* S*, see 1.3) to the category of Dieudonné crystals over* $(S/W)_{cris}$*. Furthermore,*

i) *For an object* G *in* $\mathfrak{C}_S^1$*, canonically we have* $\mathbf{D}(G) \cong \Phi^*\omega_{G/S}$ *if* $F_{G/S} = 0$, *and* $\mathbf{D}(G) \cong \Phi^*\mathcal{F}$ *if moreover* $V_{G/S} = 0$*, where* $\mathcal{F}$ *is the* α-*sheaf of* G.

ii) *For any p-divisible group* G *over* S*, there is a canonical isomorphism*

$$\mathbf{D}(G^t) \cong \mathrm{Hom}_{O_{S/W}}(\mathbf{D}(G), O_{S/W}). \tag{5.4.1}$$

iii) *The morphisms* F *and* V *of* $\mathbf{D}(G)$ *are induced by the relative Frobenius and Verschiebung of* G *respectively.*

5.5. Representability of the functor $\mathbf{D}(G)$.

Remark. If $S = \mathrm{Spec}(K)$ for a perfect field K, then for any object G in $\mathfrak{C}_S$, the crystal $\mathbf{D}(G)$ (as a functor) is represented by its value on W, which is just the Dieudonné module $D(G)$.

However, for general S, the representability of $\mathbf{D}(G)$ may not hold. But in this book we will only deal with a very good case, in which $\mathbf{D}(G)$ is uniquely determined up to isomorphism by its value on $W(R)$, where R is a domain and $S = \mathrm{Spec}(R)$. Thus we have a way to treat $\mathbf{D}(G)$ as a $W(R)[F]$-module (see 7.4). For this reason, we will also use the terminology "Dieudonné module over S" for general S, by slight abuse of language.

5.6. Supersingular Dieudonné modules over W.

Definition. Let K be a perfect field. A Dieudonné module M over $W = W(K)$ is called *supersingular* (resp. *superspecial, supergeneral*) *of genus* g if $M \otimes_W W(k) \cong D(X)$ for some supersingular (resp. superspecial, supergeneral) abelian variety X of dimension g over k.

31

For our choice of E (see 1.2), by (1.2.2) we have $F_E + V_E = 0$ on E, hence $D(E \otimes K) \cong A/A(F+V)$ by Fact 5.2.iv), here A and $A(F+V)$ (a left ideal of A) are viewed as left A-modules, and F, V are induced by F_E, V_E respectively. If $K \supset \mathsf{F}_{p^4}$, then there is an element $\epsilon \in W(K) - pW(K)$ satisfying $\epsilon^{\sigma^2} = -\epsilon$. Choose $\epsilon \cdot 1$ as a generator instead of 1, we see that $A/A(F+V)$ is isomorphic to

$$A_{1,1} = A/A(F - V). \tag{5.6.1}$$

More generally, if $K \supset \mathsf{F}_{p^4}$, then $D(E^g \otimes K) \cong A_{1,1}^{\oplus g}$, hence an isogeny $E^g \otimes K \to X$ induces a monomorphism $D(X) \hookrightarrow A_{1,1}^{\oplus g}$ which identifies $D(X)$ with a Dieudonné submodule of $A_{1,1}^{\oplus g}$ of finite colength; conversely, any Dieudonné submodule $M \subset A_{1,1}^{\oplus g}$ of finite colength gives an isogeny $\rho : E^g \otimes K \to X$ such that $D(\ker(\rho)) \cong A_{1,1}^{\oplus g}/M$. Hence by definition we have:

i) An abelian variety X over K is supersingular (resp. superspecial, supergeneral) iff $D(X)$ is so.

ii) A Dieudonné module M over W is supersingular (resp. superspecial, supergeneral) iff $M \otimes_W W(k)$ is so.

iii) An abelian variety X over K is superspecial of genus g iff $D(X \otimes_K k) \cong A_{1,1}^{\oplus g} \otimes_W W(k)$; an abelian variety X over K is supersingular of genus g iff $D(X \otimes_K k)$ is isomorphic to a Dieudonné submodule of $A_{1,1}^{\oplus g} \otimes_W W(k)$ of finite colength.

iv) A supersingular Dieudonné module M of genus g over W is superspecial iff $a(M) = g$ (see (5.2.8)).

Remark. The Dieudonné module of a supersingular abelian variety over K is a supersingular Dieudonné module; however the converse is not true: for example there is no elliptic curve E' over F_p such that $D(E') \cong A_{1,1}$. Furthermore, a supersingular Dieudonné module M of genus g over $W = W(K)$ may not be embedded into $A_{1,1}^{\oplus g}$ as a Dieudonné submodule.

For any supersingular Dieudonné module M over $W = W(K)$, there is a largest superspecial Dieudonné submodule $S_0 M \subset M$ which contains all superspecial submodules of M. (This was first studied by Manin, see [48, p. 38], also see [45, Section 1].) Dually, there is a smallest superspecial submodule $S^0 M \subset M \otimes_W (\text{q.f.}(W))$ containing M. If $S^0 X = Y \to X$ is a minimal isogeny over K (see 1.8), then the induced monomorphism $D(X) \hookrightarrow D(Y)$ induces a canonical isomorphism $D(S^0 X) \cong S^0 D(X)$. Similarly we have $S_0 D(X) \cong D(S_0 X)$ canonically.

We will use the following ([45, Section 1]):

Fact. For a supersingular Dieudonné module $M \subset A_{1,1}^{\oplus g}$ of genus g over $W = W(K)$, the following are equivalent:

i) $a(M) = 1$.

ii) $M = Av$ for some $v \in M$.

iii) $S_0 M = F^{g-1} S^0 M$.

iv) M is supergeneral.

5.7. Skeleton.

Let K be a perfect field containing F_{p^2} and $W = W(K)$. By our choice of E (see 1.2), we have $\operatorname{End}(E \otimes K) \cong \mathcal{O}$ (see (1.2.4)). Let $H = \mathcal{O} \otimes \mathsf{Z}_p$. Then $H \cong \operatorname{End}_A(A_{1,1})$, and

$$H \cong W(\mathsf{F}_{p^2})[F]/(F^2 - p, Fa - a^\sigma F | a \in W(\mathsf{F}_{p^2})) \tag{5.7.1}$$

which is the quaternion algebra over $\mathsf{Z}_p = W(\mathsf{F}_p)$ (see [45, Section 1]).

Let M be a Dieudonné submodule of $A_{1,1}^{\oplus g}$ of finite colength. The *skeleton* of M (see [45, Section 1]) is the free H-module

$$\tilde{M} = \{x \in M | Fx = Vx\}. \tag{5.7.2}$$

Clearly

$$S_0 M = \tilde{M} \otimes_{W(\mathsf{F}_{p^2})} W \subset M. \tag{5.7.3}$$

5.8. An equivalence relation.

Let K be a perfect field containing F_{p^2} and $N = A_{1,1}^{\oplus g}$. For any $v \in N$, denote by $\bar{v}$ the image of v in $N/FN \cong K^{\oplus g}$. We have ([45, Section 1]):

Fact. *Let $v \in N$ and $\bar{v} = (a_1, ..., a_g)$. Then the following are equivalent:*

i) $S^0(Av) = N$.

ii) $a_1, ..., a_g$ *are linearly independent over* F_{p^2}.

iii) $J(a_1, ..., a_g) := \begin{vmatrix} a_1 & \cdots & a_g \\ a_1^{p^2} & \cdots & a_g^{p^2} \\ \vdots & \ddots & \vdots \\ a_1^{p^{2g-2}} & \cdots & a_g^{p^{2g-2}} \end{vmatrix} \neq 0.$

In fact,

$$J(y_1, ..., y_g) = (-1)^{g(g-1)/2} \prod_{\substack{1 \leq i \leq g \\ \lambda_1, ..., \lambda_g \in \mathsf{F}_{p^2}}} (y_i + \lambda_{i+1} y_{i+1} + ... + \lambda_g y_g). \tag{5.8.1}$$

Furthermore, if K is infinite and $M \subset N$ such that $S^0 M = N$, then there exists $v \in M$ such that $S^0(Av) = N$.

5.9. Quasi-polarizations.

A polarization of an abelian variety X over a perfect field K induces a form which is non-degenerate alternating W-bilinear form $\langle, \rangle$ on $D(X^t)$ such that $\langle Fx, y \rangle = \langle x, Vy \rangle^\sigma$ (see [60, p. 101]). Such a bilinear form on a Dieudonné module will be called a *quasi-polarization*. A quasi-polarization is called *principal* if it is a perfect pairing.

Let M be a Dieudonné module which is a free W-module. Let

$$M^t = \operatorname{Hom}_W(M, W). \tag{5.9.1}$$

33

Then M^t has a Dieudonné module structure by

$$(Ff)(v) = f(Vv)^\sigma, \quad (Vf)(v) = f(Fv)^{\sigma^{-1}} \quad (\forall f \in M^t, \ v \in M). \qquad (5.9.2)$$

We call M^t the *dual* Dieudonné module of M.

Therefore a quasi-polarization of M induces a monomorphism of A-modules $M \to M^t$. We also have $D(X^t) \cong D(X)^t$.

Remark. We can also talk about "quasi-polarizations of p-divisible groups", by the anti-equivalence D (see Fact 5.2).

We will also use the terminology *quasi-polarized Dieudonné module* which means a Dieudonné module together with a quasi-polarization. For a quasi-polarized Dieudonné module M, we will identify M with its image in M^t under the induced monomorphism $M \to M^t$.

More generally, a polarization of an abelian scheme X over S also induces a nondegenerate alternating form on its Dieudonné module.

6. PFTQs of Dieudonné modules over W

In this chapter we study supersingular Dieudonné modules over $W = W(K)$, where K is a perfect field containing $\mathbf{F}_{p^2}$ (see Remark 5.6).

6.1. Quasi-polarizations of superspecial Dieudonné modules.

Proposition (compare [12, Proposition 5.1]). *Let $\{M, \langle , \rangle\}$ be a quasi-polarized superspecial Dieudonné module of genus g over $W = W(K)$. Suppose $M \cong A_{1,1}^{\oplus g}$ (e.g. when $K = k$). Then one can decompose*

$$M \cong M_1 \oplus M_2 \oplus \cdots \oplus M_d \quad (\langle M_i, M_j \rangle = 0 \text{ if } i \neq j), \tag{6.1.1}$$

where each M_i is of either of the following types:

i) *a genus 1 quasi-polarized superspecial Dieudonné module over W generated by some x such that $\langle x, Fx \rangle = p^r \epsilon$ for some $r \in \mathbf{Z}$ and $\epsilon \in W - pW$ with $\epsilon^\sigma = -\epsilon$; or*

ii) *a genus 2 quasi-polarized superspecial Dieudonné module over W generated by some x, y such that $\langle x, y \rangle = p^r$ for some $r \in \mathbf{Z}$, and $\langle x, Fx \rangle = \langle y, Fy \rangle = \langle x, Fy \rangle = \langle y, Fx \rangle = 0$.*

Proof. The bilinear form $\langle , \rangle$ can be viewed as a monomorphism $M \to M^t$, which sends the skeleton $\tilde{M}$ (see 5.7) of M to the skeleton of M^t. Hence $\langle , \rangle$ is induced by an alternating form $\tilde{M} \times \tilde{M} \to W(\mathbf{F}_{p^2})$. So we need only to prove the following:

Lemma. *Let N be a free H-module of rank g together with a non-degenerate alternating $W(\mathbf{F}_{p^2})$-bilinear form $\langle , \rangle$ such that $\langle Fx, y \rangle = \langle x, Fy \rangle^\sigma$ for all $x, y \in N$. Then one can decompose*

$$N \cong N_1 \oplus N_2 \oplus \cdots \oplus N_d \quad (\langle N_i, N_j \rangle = 0 \text{ if } i \neq j), \tag{6.1.2}$$

where each N_i is of either of the following types:

i) *a free H-module of rank 1 with basis $\{x\}$ such that $\langle x, Fx \rangle = p^r \epsilon$ for some $r \in \mathbf{Z}$ and $\epsilon \in W(\mathbf{F}_{p^2}) - pW(\mathbf{F}_{p^2})$ with $\epsilon^\sigma = -\epsilon$; or*

ii) *a free H-module of rank 2 with basis $\{x, y\}$ such that $\langle x, y \rangle = p^r$ for some $r \in \mathbf{Z}$, and $\langle x, Fx \rangle = \langle y, Fy \rangle = \langle x, Fy \rangle = \langle y, Fx \rangle = 0$.*

Proof of the lemma. First find $r \in \mathbf{Z}$ such that $\langle N, N \rangle \subset p^r W(\mathbf{F}_{p^2})$ but $\langle N, N \rangle \not\subset p^{r+1} W(\mathbf{F}_{p^2})$. Let

$$N' = \{x \in N | \langle x, N \rangle \subset p^{r+1} W(\mathbf{F}_{p^2})\}. \tag{6.1.3}$$

Then $pN \subset N'$. So $\overline{N} = N/N'$ is a linear space over $\mathbf{F}_{p^2}$, on which F is a semi-linear map and $F^2 = p \cdot$ is the zero map. Let $\overline{N}_0 = \ker(F)$. Then $F\overline{N} \subset \overline{N}_0$. Note that $\langle , \rangle$ induces an alternating $\mathbf{F}_{p^2}$-bilinear form

$$\overline{N} \times \overline{N} \to p^r W(\mathbf{F}_{p^2})/p^{r+1} W(\mathbf{F}_{p^2}) \cong \mathbf{F}_{p^2}, \tag{6.1.4}$$

which is non-degenerate by our construction. We claim that $(F\overline{N})^\perp = \overline{N}_0$. Indeed,

$$\langle \dot{x}, F\overline{N}\rangle = 0 \;\Leftrightarrow\; \langle F\dot{x}, \overline{N}\rangle = 0 \;\Leftrightarrow\; F\dot{x} = 0. \tag{6.1.5}$$

There are two possible cases:

Case 1: $\overline{N} \neq \overline{N}_0$. In this case we can find $\dot{x} \in \overline{N}$ such that $\langle \dot{x}, F\dot{x}\rangle \neq 0$. Indeed, take $\dot{x}$ such that $F\dot{x} \neq 0$, and take $\bar{y}$ such that $\langle F\dot{x}, \bar{y}\rangle \neq 0$. Then for any $\lambda \in \mathsf{F}_{p^2}$,

$$\langle \dot{x} + \lambda\bar{y}, F(\dot{x} + \lambda\bar{y})\rangle = \langle \dot{x}, F\dot{x}\rangle + \lambda\lambda^\sigma\langle \bar{y}, F\bar{y}\rangle + \lambda^\sigma\langle \dot{x}, F\bar{y}\rangle + \lambda\langle \bar{y}, F\dot{x}\rangle, \tag{6.1.6}$$

the last term of (6.1.6) equals $-(\lambda^\sigma\langle \dot{x}, F\bar{y}\rangle)^\sigma$. If $\langle \dot{x}, F\dot{x}\rangle = \langle \bar{y}, F\bar{y}\rangle = 0$, then we can take λ such that $\lambda^\sigma\langle \dot{x}, F\bar{y}\rangle \notin \mathsf{F}_p$, so $\langle x + \lambda y, F(\dot{x} + \lambda\bar{y})\rangle \neq 0$. Therefore we can get an element $x \in N$ such that $\langle x, Fx\rangle \notin p^{r+1}W(\mathsf{F}_{p^2})$. Clearly $\langle x, Fx\rangle = p^r\epsilon$ for some $\epsilon \in W(\mathsf{F}_{p^2}) - pW(\mathsf{F}_{p^2})$ with $\epsilon^\sigma = -\epsilon$. Note that if $\epsilon' \in W(\mathsf{F}_{p^2}) - pW(\mathsf{F}_{p^2})$ satisfies $\epsilon'^\sigma = -\epsilon'$, then $\epsilon' = c\epsilon$ for some $c \in W(\mathsf{F}_p) - pW(\mathsf{F}_p)$. We can take $\alpha = a + bF \in H$ such that

$$c = |\alpha| = aa^\sigma - pbb^\sigma = (a + bF)(a^\sigma - bF) \tag{6.1.7}$$

because $|\ |: H \to W(\mathsf{F}_p)$ is surjective (see [45, Lemma 1.2]), hence $\langle \alpha x, F\alpha x\rangle = p^r\epsilon'$. Thus we can fix ϵ before choosing x. Let $N_1 = Hx$. It is easy to check that $N \cong N_1 \oplus N_1^\perp$.

Case 2: $\overline{N} = \overline{N}_0$ (i.e. $\langle FN, N\rangle \subset p^{r+1}W(\mathsf{F}_{p^2})$). Since $\overline{N}$ has a non-degenerate alternating form, $\dim_{\mathsf{F}_{p^2}} \overline{N} \geq 2$. Take $\bar{x}, \bar{y} \in \overline{N}$ such that $\langle \bar{x}, \bar{y}\rangle \neq 0$ and lift them to some $x, y \in N$. We may assume that $\langle Fx, y\rangle = 0$, otherwise we can take $y - \lambda Fy$ instead of y, where $\lambda = \langle Fx, y\rangle/p\langle x, y\rangle^\sigma$. Let v be the p-adic valuation. If $\langle x, Fx\rangle \neq 0$ and $\langle y, Fy\rangle \neq 0$, say $v(\langle x, Fx\rangle) \geq v(\langle y, Fy\rangle)$, then for any $x' = x + \alpha y$ ($\alpha \in H$), it is easy to calculate that $\langle x', Fx'\rangle = \langle x, Fx\rangle + |\alpha|\langle y, Fy\rangle$. Since $\langle x, Fx\rangle/\langle y, Fy\rangle \in W(\mathsf{F}_p)$ and $|\ |: H \to W(\mathsf{F}_p)$ is surjective (see Case 1 above), we can take α such that $\langle x', Fx'\rangle = 0$. Take x' instead of x. As above, we can take y such that $\langle y, Fx\rangle = 0$. If $\langle y, Fy\rangle \neq 0$, take $y' = y + \beta Fx$ ($\beta \in W(\mathsf{F}_{p^2})$) instead of y and calculate that

$$\langle y', Fy'\rangle = \langle y, Fy\rangle + p(\beta\langle x, y\rangle^\sigma - \beta^\sigma\langle x, y\rangle). \tag{6.1.8}$$

Again we can take β such that $\langle y', Fy'\rangle = 0$. Therefore we can get $x, y \in N - N'$ such that $\langle x, Fx\rangle = \langle y, Fy\rangle = \langle x, Fy\rangle = \langle y, Fx\rangle = 0$ and $\langle x, y\rangle \notin p^{r+1}W(\mathsf{F}_{p^2})$. By taking $p^r\langle x, y\rangle^{-1}y$ instead of y we also have $\langle x, y\rangle = p^r$. Let $N_1 = Hx + Hy$. Again one checks that $N \cong N_1 \oplus N_1^\perp$.

The lemma is then proved by induction.   Q.E.D.

Remark. In Proposition 6.1, if $g = 2$, then M is either of type ii) or isomorphic to a direct sum of two submodules of type i) (but possibly corresponding to different r's). In the first case, the quasi-polarization gives $M = F^{2r+1}M^t$. In the second case, if the two submodules of type i) have the same r, then $M = F^{2r}M^t$. If $M = F^nM^t$, one can find generators $\{x, y\}$ of M^t such that $\langle x, Fx\rangle = \langle y, Fy\rangle = \langle x, F^ny\rangle = 0$ and $\langle x, F^{n+1}y\rangle = 1$.

Corollary. *Up to isomorphism, there is exactly one quasi-polarized superspecial Dieudonné module M of genus g over $W(k)$ such that the quasi-polarization gives $M \xrightarrow{\sim} F^{g-1}M^t \subset M^t$ (cf. 5.6).*

6.2. PFTQ of Dieudonné modules.

We note that if $Y_{g-1} \supset \ldots \supset Y_0$ is a rigid PFTQ with respect to some η over a perfect field $K \supset \mathbb{F}_{p^4}$, then $M_i = D(Y_i)$ $(0 \le i \le g-1)$ form a filtration of $M_{g-1} \supset M_0$ such that

    i) $M_{g-1} \cong A_{1,1}^{\oplus g}$ with $M_{g-1}^t = F^{g-1} M_{g-1}$;

    ii) $(F,V) M_i \subset M_{i-1}$ and $\dim_K(M_i/M_{i-1}) = i$ $(1 \le i \le g-1)$;

    iii) $F^{i-j} V^j M_i \subset M^i$ $(\forall 0 \le j \le i/2)$, where $M^i = M_i^t$;

    iv) $M_i = M_0 + F^{g-1-i} M_{g-1}$ $(1 \le i \le g-1)$.

This motivates us to set up the following definition.

Definition. A *PFTQ of Dieudonné modules* of genus g over W is a filtration $M_{g-1} \supset \ldots \supset M_0$ of quasi-polarized Dieudonné modules over W satisfying i-iii) above. It is called *rigid* if iv) above also holds.

6.3. Existence of rigid PFTQ for a quasi-polarized supersingular Dieudonné module.

Proposition. *Let M be a principally quasi-polarized supersingular Dieudonné module of genus g over $W = W(k)$. Then there is a rigid PFTQ of Dieudonné modules $\{M_{g-1} \supset \ldots \supset M_0\}$ over W such that $M_0 \cong M$ as quasi-polarized Dieudonné modules.*

Proof. We use induction on g. Let $N = S^0 M$. By Proposition 6.1 and Remark 6.1, one can decompose $N = N' \oplus N''$, where N', N'' are superspecial, $\langle N', N'' \rangle = 0$, $g(N') = 2$, and $N''^t = F^n N'$ for some $n < g$. Take generators $\{x, y\}$ of N' as in Remark 6.1. Then the projection $M \to Ax$ is an epimorphism whose dual is $AF^n y \hookrightarrow M$. (Note that $AF^n y = Ay \cap M$ since $S_0 M = S_0(M^t) = N^t$.) We then get a complex of A-modules

$$C^{\cdot} \; : \quad 0 \to AF^n y \to M \to Ax \to 0 \tag{6.3.1}$$

which is self-dual. Let $M' = H^1(C^{\cdot})$. Then $g(M') = g - 2$ and $\langle \, , \, \rangle$ induces a principal quasi-polarization on M'. By induction, there is a rigid PFTQ of Dieudonné modules $\{M_i' | 0 \le i < g-2\}$ such that $M_0' \cong M'$. Note that M'^{g-3} can be viewed as an A-submodule of N''. Let M_{g-1} be the A-submodule of $M \otimes_W (\text{q.f.}(W))$ generated by $x, F^{n-g+1} y$ and $F^{-1} M_{g-3}'$. Then $M^{g-1} = M_{g-1}^t$ is generated by $F^{g-1} x, F^n y$ and FM'^{g-3}. Hence $M^{g-1} = F^{g-1} M_{g-1}$.

    We first check that $M \subset M_{g-1}$. By Fact 5.8, we can choose $v = x + (a+bF)y + w \in M$ (where $a, b \in W$, $w \in N''$) such that $S^0(Av) = N$. This is equivalent to that $F^i V^{g-1-i} \bar{v}$ $(0 \le i < g)$ are linearly independent over k. Hence $F^i V^{g-2-i}(F-V)\bar{v}$ $(0 \le i \le g-2)$ are linearly independent over k, and this shows that $S^0((F-V)v) = AFy \oplus FN''$. Since $(F-V)v \in \ker(M \to Ax)$, we get that $FN'' \subset M_{g-3}'$. Hence $M \subset N \subset N' \oplus F^{-1} M_{g-3}' \subset M_{g-1}$.

    Let $M_0 = M$ and define M_i $(0 < i < g)$ by iv) in 6.2. Then i) in 6.2 already holds. To check iii), note that $M^i = M_i^t = F^i M_{g-1} \cap M$. For ii), we need only check the second part. Since $F''y \in M^i$ $(0 \leq i < g)$, we have

$$
\begin{aligned}
\dim_k(M_{i+1}/M_i) &= \dim_k(AF^{g-2-i}x/AF^{g-1-i}x) + \dim_k(M'_i/M'_{i-1}) \\
&= i+1 \quad (0 \leq i \leq g-2).
\end{aligned}
\tag{6.3.2}
$$

Q.E.D.

Remark. In the above proof, $\{M'_i | 0 \leq i < g-2\}$ and $\{M_i | 0 \leq i < g\}$ satisfy the following conditions:

$$
((Ay \dotplus F^{-1}M'_{g-3}) \cap M_i)/AF^{g-1-i}y \cong M'_{i-1} \ (1 \leq i \leq g-2).
\tag{6.3.3}
$$

and

$$
((Ay \dotplus F^{-1}M'_{g-3}) \cap (F,V)M_i) \subset AFy \dotplus M'_{i-2} \ (2 \leq i \leq g-2).
\tag{6.3.4}
$$

since $(F,V)M'_{i-1} \subset M'_{i-2}$.

6.4. The number of PFTQs for a quasi-polarized supersingular Dieudonné module.

Remark. If M is supergeneral, then the PFTQ of Dieudonné modules in Proposition 6.3 is unique up to isomorphism, because it is determined by M and $S^0 M$. If M is not supergeneral, then we can at least guarantee that there are only a finite number of *rigid* PFTQs of Dieudonné modules ending at M up to isomorphism. Indeed, such a PFTQ of Dieudonné modules is uniquely determined by $M_{g-1} \supset M$. If N is the skeleton of M and N_{g-1} is the skeleton of M_{g-1}, then by 5.7 we see M_{g-1} can be uniquely determined by $N_{g-1}/N \subset F^{1-g}N/N \cong (H/F^{g-1}H)^{\oplus g}$ which is a finite set.

However, when M is not supergeneral, we may have infinitely many isomorphism classes of non-rigid PFTQs of Dieudonné modules $\{M_{g-1} \supset ... \supset M_0\}$ such that $M_0 \cong M$ (as quasi-polarized Dieudonné modules). For example, when $g = 3$, let M be a direct sum of three copies of quasi-polarized Dieudonné modules of type i) in Proposition 6.1 with $r = 0$. Let $M_2 = F^{-1}M$ and $M_0 = M$. To give an M_1 is equivalent to giving a vector $(a_1, a_2, a_3) \in M_2/M \cong K^{\oplus 3}$ such that $a_1^{p^2+1} + a_2^{p^2+1} + a_3^{p^2+1} = 0$. Therefore we get a one-parameter family of PFTQs of Dieudonné modules ending at M. Note that $\mathrm{Aut}_A(M_2/M)$ is finite, hence there are infinitely many PFTQs of Dieudonné modules ending at M up to isomorphism (cf. Example 3.3).

7. Moduli of rigid PFTQs of Dieudonné modules

In this chapter we define PFTQs of Dieudonné modules over a reduced K-scheme S, and show that the category of rigid PFTQs of Dieudonné modules of genus g has a fine moduli scheme which is integral. In this way we can prove the weak form of Proposition 4.3 (see 4.11).

Throughout this chapter we assume $K \supset F_{p^4}$ unless otherwise specified (see 5.6 for the reason), though some of the statements hold over F_{p^2}.

7.1. The Dieudonné module of a PFTQ over a general base.

Let K be a perfect field and $W = W(K)$. Then $D(E^g \otimes K) \cong A_{1,1}^{\oplus g}$ over W (see (5.6.1)). We fix a quasi-polarization $\langle,\rangle$ of $N = A_{1,1}^{\oplus g}$ such that $N^t = F^{g-1}N$. Note that $\{N, \langle,\rangle\}$ is unique up to isomorphism by Corollary 6.1.

Let $\{Y_i, \eta_i(0 \le i < g); \rho_i(0 < i < g)\}$ be a PFTQ with respect to some η over a *reduced* K-scheme $S = \mathrm{Spec}(R)$. Then the Dieudonné crystal functor $\mathbf{D}$ (see Fact 5.4) gives locally free $W(R)$-modules $M_i = \mathbf{D}(Y_i)_{W(R)}$ (of rank $2g$) which form a filtration $\{M_{g-1} \supset ... \supset M_0\}$ together with a $W(R)$-bilinear form satisfying

i) $M_{g-1} \cong N \otimes_W W(R)$, compatible with the bilinear forms;

ii) $FM_i^{(p)} \subset M_{i-1}$, $VM_i \subset M_{i-1}^{(p)}$, and $(M_i/M_{i-1}) \otimes_{W(R)} R$ is a locally free R-module of rank i $(1 \le i \le g - 1)$ (in fact $M_i/M_{i-1} \cong \mathbf{D}(G)_{W(R)}$, where $G = \ker(\rho_i : Y_i \to Y_{i-1})$, hence $(M_i/M_{i-1}) \otimes_{W(R)} R \cong \mathbf{D}(G)_R \cong \mathcal{F}^{(p)}(S)$, where $\mathcal{F}$ is the α-sheaf of G);

iii) $F^{i-j} \circ V^j M_i^{(p^{i-2j})} \subset M^i$ $(\forall 0 \le j \le i/2)$, where $M^i = M_i^t \subset M_i$ (the dual with respect to the bilinear form).

If the PFTQ is rigid, then we also have

iv) $M_i = M_0 + F^{g-1-i}M_{g-1}^{(p^{g-1-i})}$ $(1 \le i \le g - 1)$.

7.2. Quasi-polarized Dieudonné module over a general base.

The observation in 7.1 motivates us to set up the following definitions. We take N as in 7.1.

Definition. Let $S = \mathrm{Spec}(R)$ be a reduced K-scheme. A *quasi-polarized Dieudonné module* of genus g over S is a locally free $W(R)$-module M of rank $2g$ together with $W(R)$-linear maps $F : M^{(p)} \to M$ and $V : M \to M^{(p)}$ such that

$$F \circ V = p \cdot \mathrm{id}_M, \quad V \circ F = p \cdot \mathrm{id}_{M^{(p)}} \tag{7.2.1}$$

and a $W(R)$-bilinear alternating form $\langle,\rangle : M^t \otimes_{W(R)} M^t \to W(R)$ such that

$$\langle Fv, w\rangle = \langle v, Vw\rangle^\sigma \quad \forall v \in M^{t(p)}, \ w \in M^t, \tag{7.2.2}$$

39

where $M^t = \text{Hom}_{W(R)}(M, W(R))$.

Note that $\langle,\rangle$ induces a monomorphism $M^t \hookrightarrow M$. We will identify M^t with a $W(R)$-submodule of M via this monomorphism. Thus $\langle,\rangle$ induces a $W(R)$-bilinear form (also denoted by $\langle,\rangle$) $M \otimes_{W(R)} M \to W(R) \otimes \mathbf{Q}$.

7.3. PFTQ of Dieudonné modules.

Definition. Let $W = W(K)$ and $N, \langle,\rangle$ be as in 7.1. A *PFTQ of Dieudonné modules* of genus g over a reduced K-scheme $S = \text{Spec}(R)$ is a filtration of quasi-polarized Dieudonné modules $\{M_{g-1} \supset ... \supset M_0\}$ of genus g over S such that

   i) $M_{g-1} \cong N \otimes_W W(R)$ as quasi-polarized Dieudonné modules over $W(R)$;
   ii) $FM_i^{(p)} \subset M_{i-1}$, $VM_i \subset M_{i-1}^{(p)}$, and $(M_i/M_{i-1})_{W(R)}R$ is a locally free R-module of rank i ($1 \le i \le g - 1$);
   iii) $F^{i-j} \circ V^j M_i^{(p^{i-2j})} \subset M^i$ ($\forall 0 \le j \le i/2$), where $M^i = M_i^t \subset M_i$,

which is called *rigid* if in addition that

   iv) $M_i = M_0 + F^{g-1-i} M_{g-1}^{(p^{g-1-i})}$ ($1 \le i \le g - 1$).

The above two definitions can be generalized to over any K-scheme S of finite type. Note that $\{M_{g-1} \supset ... \supset M_0\}$ is determined by the induced sequence of homomorphisms $M_0 \otimes_{W(R)} W_n(R) \to ... \to M_{g-1} \otimes_{W(R)} W_n(R)$ for any $n \gg 0$.

7.4. The correspondence between PFTQs and PFTQs of Dieudonné modules.

Let $S = \text{Spec}(R)$ be a reduced K-scheme and $\{Y_i, \eta_i(0 \le i < g); \rho_i(0 < i < g)\}$ be a PFTQ with respect to some η over S. Then by 7.1 we have an induced PFTQ of Dieudonné modules $\{M_{g-1} \supset ... \supset M_0\}$ of genus g over S.

Lemma. *The PFTQ* $\{Y_i, \eta_i(0 \le i < g); \rho_i(0 < i < g)\}$ *is uniquely determined up to isomorphism by* η *and the PFTQ of Dieudonné modules* $\{M_{g-1} \supset ... \supset M_0\}$. *Furthermore, when* $S = \text{Spec}(K)$, *the functor* D *(see 5.2) gives a natural bijection:*

   $D :$ {PFTQs with respect to η over K}/isomorphisms
   $\longleftrightarrow$ {PFTQs of Dieudonné modules of genus g over $W = W(K)$}

in which rigid PFTQs correspond to rigid PFTQs of Dieudonné modules.

Proof. Given η and $M_{g-1} \supset ... \supset M_0$, we use inverse induction to recover the Y_i's. First, $\{Y_{g-1}, \eta_{g-1}\}$ is given by η. To get Y_{i-1} from Y_i, it is enough to determine the subgroup scheme $G = \text{ker}(\rho_i)$ as a subgroup scheme of $Y_i[F]$. This is then equivalent to determining the α-sheaf $\mathcal{F}$ of G (see 2.4) as a quotient of $\omega_{Y_i[F]/S} \cong \omega_{Y_i/S}$, because $\omega_{Y_i[F]/S} \twoheadrightarrow \mathcal{F}$ determines G^D as a quotient group scheme of $Y_i[F]^D \cong \text{ker}(V_{Y_i/S})$. By Fact 5.4.i) we have

$$\omega_{G^{(p)}/R} \cong \mathcal{F}^{(p)}(S) \cong (M_i/M_{i-1}) \otimes_{W(R)} R, \tag{7.4.1}$$

also we have

$$\omega_{Y_i[F]/R}^{(p)} \cong (M_i/FM_i^{(p)}) \otimes_{W(R)} R. \tag{7.4.2}$$

Hence $\omega^{(p)}_{Y_i[F]/R} \to \mathcal{F}^{(p)}(S)$ is determined by $M_{i-1} \hookrightarrow M_i$. Finally, since $\mathcal{F}$ is flat over R which is a domain, we see $\omega_{Y_i[F]/R} \to \mathcal{F}(S)$ is determined by $\omega^{(p)}_{Y_i[F]/R} \to \mathcal{F}^{(p)}(S)$.

When $S = \mathrm{Spec}(K)$, we can construct a PFTQ from any PFTQ of Dieudonné modules by the above procedure.   Q.E.D.

7.5. Construct a PFTQ from a PFTQ of Dieudonné modules.

Remark. Given η and a PFTQ of Dieudonné modules

$$\{M_{g-1} \supset ... \supset M_0\} \tag{7.5.1}$$

over a reduced K-scheme $S = \mathrm{Spec}(R)$, one can use the procedure in the proof of Lemma 7.4 to construct a PFTQ

$$\{Y_i, \eta_i(0 \le i < g); \rho_i(0 < i < g)\} \tag{7.5.2}$$

(with respect to η) whose corresponding PFTQ of Dieudonné modules is isomorphic to

$$\{M^{(p^{g-1})}_{g-1} \supset ... \supset M^{(p^{g-1})}_0\}. \tag{7.5.3}$$

7.6. Moduli of PFTQs of Dieudonné modules.

7.6. Let $\mathfrak{N}_g$ be the category of *rigid* PFTQs of Dieudonné modules of genus g over reduced $\mathbf{F}_{p^4}$-schemes. The aim of this chapter is to prove:

Proposition. *The category $\mathfrak{N}_g$ has a fine moduli scheme $\mathcal{N}_g$, which is smooth over $\mathbf{F}_{p^4}$ and geometrically integral of dimension $[g^2/4]$. Furthermore, if $\{M_{g-1} \supset ... \supset M_0\}$ is the universal PFTQ of Dieudonné modules over $\mathcal{N}_g$, then the generic fiber of M_0 has a-number 1.*

We will use double induction to do this, as mentioned in 4.11. By Example 3.8, the statements hold when $g \le 2$. Let $g \ge 3$ and assume the claims hold for $\mathfrak{N}_{g-2}$.

Let $\{M_{g-1} \supset ... \supset M_0\}$ be a rigid PFTQ of Dieudonné modules over some $\mathbf{F}_{p^4}$-scheme S. By Lemma 6.1 we can choose an H-basis $\Theta := \{x, y, x_1, ..., x_{g-2}\}$ of the skeleton of $N = A^{\oplus g}_{1,1}$ such that

$$\langle x, F^g y \rangle = 1, \langle x, Fx \rangle = \langle y, Fy \rangle = \langle x, F^{g-1}y \rangle = 0, \langle x, L \rangle = \langle y, L \rangle = 0, \tag{7.6.1}$$

where L is the H-submodule of N generated by $x_1, ..., x_{g-2}$. For convenience we choose $x_1, ..., x_{g-2}$ such that when g is even

$$\begin{aligned} &\langle x_i, F^g x_{g-1-i} \rangle = 1 \ (\forall i), \langle x_i, F^{g-1} x_j \rangle = 0 \ (\forall i, j) \\ &\langle x_i, F^g x_j \rangle = 0 \ (\forall j \ne g-1-i) \end{aligned} \tag{7.6.2}$$

and when g is odd

$$\langle x_i, F^g x_j \rangle = \epsilon \delta_{ij}, \ \langle x_i, F^{g-1} x_j \rangle = 0 \ (\forall i, j) \tag{7.6.3}$$

41

where $\epsilon \in W(\mathbf{F}_{p^2}) - pW(\mathbf{F}_{p^2})$, $\epsilon^\sigma = -\epsilon$.

For any choice Θ of such a basis, there is a largest open subset $U \subset S$ such that the projection $M_0|_U \to W(U)Hx$ is an epimorphism. Let $A_U = W(U)H$. Since $M_0^t \cong M_0$, by taking the dual, we have a monomorphism $A_U F^{g-1}y \hookrightarrow M_0|_U$. For each i we have a complex

$$C_i : \quad A_U F^{g-1-i}y \hookrightarrow M_i|_U \twoheadrightarrow A_U x . \tag{7.6.4}$$

Let $M_i' = H^1(C_{i+1})$ $(0 \le i \le g-3)$. Then $\{M_{g-3}' \supset ... \supset M_0'\}$ is a rigid PFTQ of Dieudonné modules of genus $g-2$. Let $U^\Theta \subset U$ be the largest open subset over which M_0' has generators of the form

$$v_i' = \sum_{j=i}^{g-2} \alpha_{ij} F^i x_j \quad (\alpha_{ij} \in A_U, \alpha_{ii} = 1) \quad (1 \le i \le g-2). \tag{7.6.5}$$

Clearly $S = \bigcup_\Theta U^\Theta$.

We fix a choice of Θ until 7.13. Denote by $\mathfrak{N}^\Theta$ the subcategory of $\mathfrak{N}_g$ consisting of objects $\{M_{g-1} \supset ... \supset M_0\}$ such that $M_0 \to W(S)Hx$ is an epimorphism and M_0' locally has generators of the form (7.6.5).

Let $\mathfrak{U}_m^\Theta$ be the category of pairs of filtrations of Dieudonné modules $\{M_{g-3}' \supset ... \supset M_0'; M_{g-1} \supset ... \supset M_m\}$ over some reduced $\mathbf{F}_{p^4}$-scheme S such that

a) $\{M_{g-3}' \supset ... \supset M_0'\}$ is a rigid PFTQ of Dieudonné modules of genus $g-2$, and M_0' locally has generators of the form (7.6.5);

b) $\{M_{g-1} \supset ... \supset M_m\}$ is a filtration of quasi-polarized Dieudonné modules such that

    b1) $M_{g-1} \cong N \otimes_W W(S)$ as quasi-polarized Dieudonné modules over $W(S)$;

    b2) $FM_i^{(p)} \subset M_{i-1}$, $VM_i \subset M_{i-1}^{(p)}$, and $(M_i/M_{i-1}) \otimes_{W(S)} O_S$ is a locally free O_S-module of rank i $(m < i < g)$;

    b3) $F^{i-j} \circ V^j M_i^{(p^{i-2j})} \subset M^i$ $(\forall 0 \le j \le i/2)$, where $M^i = M_i^t \subset M_i$ $(m \le i < g)$;

    b4) $M_i = M_m + F^{g-1-i} M_{g-1}^{(p^{g-1-i})}$ $(m < i < g)$.

c) The projection $M_m \to W(S)Hx$ is an epimorphism and there are isomorphisms $M_{i-1}' \cong (M_i \cap M)/W(S)HF^{g-1-i}y$ $(\max(m,1) \le i \le g-2)$ compatible with the filtrations, where M is the Dieudonné submodule of M_{g-1} generated by $x_1, ..., x_{g-2}, y$;

d) $((F(M_i^{(p^2)}) + V(M_i)) \cap M)/W(S)HF^{g-i}y \subset M_{i-2}'^{(p)}$ $(\max(m,2) \le i \le g-2)$ under the isomorphisms in c).

By induction hypothesis (on $\mathfrak{N}_{g-2}$), we see $\mathfrak{U}_{g-1}^\Theta$ has a fine moduli scheme $\mathcal{U}_{g-1}^\Theta$ which is an open subscheme of $\mathcal{N}_{g-2}$. We will use inverse induction to show that each $\mathfrak{U}_m^\Theta$ has a fine moduli scheme $\mathcal{U}_m^\Theta$, hence $\mathfrak{N}_g$ has a fine moduli scheme $\mathcal{N}_g$ which has an open cover $\{\mathcal{U}_0^\Theta | \forall \Theta\}$. Since $\mathfrak{U}_m^\Theta$ (resp. $\mathcal{U}_m^\Theta$) are isomorphic to each other for different choices of Θ, we will denote them simply by $\mathfrak{U}_m$ (resp. $\mathcal{U}_m$) when we fix a Θ. Note that the truncation functors $t_m : \mathfrak{U}_{m-1} \to \mathfrak{U}_m$ give canonical morphisms $\mathcal{U}_{m-1} \to \mathcal{U}_m$ $(1 \le m \le g-1)$.

7.7. Artin-Schreier extension.

We will use the following classical result. For a field L of characteristic $p > 0$, let $\mathcal{P}L = \{x^p - x | x \in L\}$. We have (cf. [30, p.293])

Fact (Artin-Schreier). *If L'/L is a finite abelian extension such that $\mathrm{Gal}(L'/L)^p = \{1\}$, then $[(\mathcal{P}L' \cap L) : \mathcal{P}L] = [L' : L]$.*

7.8. A corollary on domains.

Corollary. *Let R be an $\mathbf{F}_{p^2}$-algebra which is a domain. Let $a_1, \ldots, a_n$, $b_1, \ldots, b_n \in R$ such that $a_1, \ldots, a_n$ are linearly independent over $\mathbf{F}_{p^2}$. Then*

$$R[t, x_1, \ldots, x_n]/(x_i^{p^2} - x_i - a_i t - b_i | 1 \leq i \leq n)$$

is a domain.

Proof. By taking the quotient field, we may assume that R is a field. We use induction on n. It is enough to show that for each i ($1 \leq i \leq n$), the polynomial $x_i^{p^2} - x_i - a_i t - b_i$ is irreducible over $L = R(t)[x_1, \ldots, x_{i-1}]/(x_j^{p^2} - x_j - a_j t - b_j | 1 \leq j < i)$. Note that if θ is a zero of $x_i^{p^2} - x_i - a_i t - b_i$, then $x_i^{p^2} - x_i - a_i t - b_i$ splits completely over $L(\theta)$. By Lemma 7.7, we have

$$p^{2i} \geq [L(\theta) : R(t)] = [(\mathcal{P}L(\theta) \cap R(t)) : \mathcal{P}R(t)] \geq p^{2i}. \tag{7.8.1}$$

This shows that $[L(\theta) : L] = p^2$, hence $x_i^{p^2} - x_i - a_i t - b_i$ is irreducible over L. Q.E.D.

7.9. Fibers of the truncation morphisms.

Let $\{M'_{g-3} \supset ... \supset M'_0; M_{g-1} \supset ... \supset M_m\}$ be an object of $\mathfrak{U}_m$ over $S = \mathrm{Spec}(R)$. Let $A_R = W(R) \otimes_W H$. Again denote by M the Dieudonné submodule of M_{g-1} generated by $\{y, x_1, ..., x_{g-2}\}$. For any $a, b \in W(R)$, denote by $\overline{a + bF}$ the image of a in R. By slight abuse of notation, for any vector v of M_i, we will denote its image in $M_i^{(p)}$ simply by v.

Lemma. *Fix a vector $v_m \in M_m$ with x-coefficient 1. Then there is a one to one correspondence*

$$\{\text{objects in } t_m^{-1}(\{M'_{g-3} \supset ... \supset M'_0; M_{g-1} \supset ... \supset M_m\})\} \longleftrightarrow$$
$$\{\text{vectors } v = v_m + \beta_{g-m}F^{g-m-1}x_{g-m} + ... + \beta_{g-2}F^{g-m-1}x_{g-2} + \beta F^{g-m-1}y$$
$$\text{satisfying A) and B) (mod } F^{g-m}M_{g-1})\}$$

where conditions A) and B) are

A) *$(F - V)v$ (mod $A_R y$) is contained in $M'^{(p)}_{m-3}$ (for $m > 2$ only);*
B) *$\langle v, (F, V)^{m-1}v \rangle \subset W(R)$.*

43

Proof. For any choice of M_{m-1}, we can find $v' \in M_{m-1}$ with x-coefficient 1. Then $v' - v_m \in M_m \cap M$. By conditions b3), c) and our assumption, there exists $w \in M_{m-1}$ and $\beta_{g-m}, ..., \beta_{g-2}, \beta \in A_R$ such that

$$v' - v_m = w + \beta_{g-m} F^{g-m-1} x_{g-m} + ... + \beta_{g-2} F^{g-m-1} x_{g-2} + \beta F^{g-m-1} y . \quad (7.9.1)$$

Take $v = v' - w$. Clearly v satisfies A) and B) by conditions b3) and d). Furthermore, any two choices of such $v \in M_{m-1}$ differ by an element in $F^{g-m} M_{g-1}$.

We now show that M^{m-1} (and hence M_{m-1}) is uniquely determined by v. Note that $M^{m-1} = M_{m-1}^t$ is generated by $F^{m-1} v$ and $M^{m-1} \cap M$. Since

$$M^{m-1} \cap M / A_R F^{g-1} y \cong M'^{m-2}, \quad (7.9.2)$$

we have an isomorphism

$$M^{m-1} \cap M / M^m \cap M \cong M'^{m-2} / M'^{m-1}. \quad (7.9.3)$$

Hence to determine M^{m-1} is equivalent to determining the lifting of each $w' \in M'^{m-2}$ in $M^{m-1} \cap M$ modulo $M^m \cap M$. Take any lifting $w \in M_m \cap M$ of w' such that

$$Fw \in M^m, \ Vw \in (M^m)^{(p)}. \quad (7.9.4)$$

Any two choices of w differ by an element in $A_R F^{g-2} y$. Hence the problem is to determine an $\alpha \in A_R$ (modulo $F A_R$) such that $w + \alpha F^{g-2} y \in M^{m-1}$. Since $v \in M_{m-1}$ we have

$$\langle v, w + \alpha F^{g-2} y \rangle \in W(R) \quad (7.9.5)$$

or equivalently

$$\overline{p \langle v, w + \alpha F^{g-2} y \rangle} = 0. \quad (7.9.6)$$

Since

$$\langle v, w + \alpha F^{g-2} y \rangle = \langle v, w \rangle + \langle v, \alpha F^{g-2} y \rangle = \langle v, w \rangle + p^{-1} \alpha, \quad (7.9.7)$$

(7.9.4) does uniquely determine $\bar{\alpha}$. Hence M^{m-1} is uniquely determined by v.

Suppose we are given a $v \in M_m$ satisfying A) and B). To find an M_{m-1}, we first note that condition c) gives an epimorphism $M^m \cap M \twoheadrightarrow M'^{m-1}$, which leads to an epimorphism

$$\psi : p^{-1} F M^{m(p)} \cap M \twoheadrightarrow p^{-1} F (M'^{m-1})^{(p)}. \quad (7.9.8)$$

Let $M' = \psi^{-1}(M'^{m-2})$. Since the dual of $p^{-1} F M^{m(p)}$ is $F M_m^{(p)}$ we have

$$\langle M', F M_m^{(p)} \rangle \subset W(R). \quad (7.9.9)$$

We now show that $M'^{(p)} \subset p^{-1} V M^m$, or equivalently

$$\langle M'^{(p)}, V M_m \rangle \subset W(R). \quad (7.9.10)$$

Let $M'' = F(M_m^{(p^2)}) + V(M_m)$. Then M'' is generated by Fv and $M'' \cap M$. By condition d) we have

$$\langle M'^{(p)}, M'' \cap M \rangle \subset W(R) \quad (7.9.11)$$

44

and (7.9.10) is a consequence of (7.9.9) and (7.9.11).

This shows that any $w' \in M'^{m-2}$ has a lifting $w \in M_m$ satisfying (7.9.4), and which can be fixed modulo M^m by (7.9.5). Let M^{m-1} be the A_R-submodule of M_m generated by M^m, $F^{m-1}v$ and all of such liftings of $w' \in M'^{m-2}$. Then B) gives $\langle v, M^{m-1} \rangle \subset W(R)$. We also have $\langle M^{m-1} \cap M, M^{m-1} \cap M \rangle \subset W(R)$ (because $\langle M'^{m-2}, M'^{m-2} \rangle \subset W(R)$). Hence $\langle M^{m-1}, M^{m-1} \rangle \subset W(R)$. Therefore

$$v \in M_{m-1} := (M^{m-1})^t \supset M^{m-1}. \tag{7.9.12}$$

Let us check conditions b), c) and d) for M_{m-1}. It is easy to see that b1), b2) and b4) are satisfied. For b3), note that by condition d) for M_m, for each $i > 0$ we have

$$(F^{m-1} - F^{m-1-i}V^i)v \pmod{A_R y} \in M'^{m-2}. \tag{7.9.13}$$

Hence B) shows $(F^{m-1} - F^{m-1-i}V^i)v \in M^{m-1}$ by our construction. Therefore $(F,V)^{m-1}v \subset M^{m-1}$, or

$$\langle M_{m-1}, (F,V)^{m-1}v \rangle \subset W(R). \tag{7.9.14}$$

By symmetry we also have (for simplicity we omit the exponent (p^i))

$$\langle v, (F,V)^{m-1}M_{m-1} \rangle \subset W(R). \tag{7.9.15}$$

To show that $(F,V)^{m-1}M_{m-1} \subset M^{m-1}$, or equivalently

$$\langle M_{m-1}, (F,V)^{m-1}M_{m-1} \rangle \subset W(R), \tag{7.9.16}$$

it remains to be checked that

$$\langle M_{m-1} \cap M, (F,V)^{m-1}M_{m-1} \cap M \rangle \subset W(R). \tag{7.9.17}$$

This is clear since

$$(F,V)^{m-1}M_{m-1} \cap M \subset (F,V)^{m-2}(M_{m-1} \cap M) \tag{7.9.18}$$

and $\langle M'_{m-2}, (F,V)^{m-2}M'_{m-2} \rangle \subset W(R)$. Therefore b) is checked. Finally, c) holds by our construction and d) is guaranteed by A). Q.E.D.

Note that B) is trivial when $m = 1$.

7.10. An equivalent condition on fibers of the truncation morphisms.

Lemma. When $m > 2$, the two conditions A) and B) are equivalent to: A) and
  B') $\langle v, p^s F^{m-1-2s}v \rangle \in W(R)$, where $s = [m/2] - 1$.

Proof. Indeed, since B') is a special case of B), we need only deduce B) from A) and B'). Condition A) gives that

$$\langle (F-V)v, (F,V)^{m-3}(F-V)v \rangle \subset W(R) \tag{7.10.1}$$

45

There are two possible cases:

Case 1: $m - 1 - 2s = 1$. Then B') gives $\langle Vv, p^s v \rangle^\sigma = \langle v, p^s Fv \rangle \in W(R)$, hence $\langle (F - V)v, p^s v \rangle \in W(R)$. (7.10.1) shows that

$$\langle (F - V)v, F^j V^{m-2-j}v \rangle - \langle (F - V)v, F^{j-1}V^{m-1-j}v \rangle \in W(R) \quad (0 < j < m).$$
(7.10.2)

By induction we get

$$\langle (F - V)v, F^j V^{m-2-j}v \rangle \in W(R) \quad (0 \le j < m).$$
(7.10.3)

Then by induction from B') we get B).

Case 2: $m - 1 - 2s = 2$. Then B') gives that

$$\langle Vv, p^s Fv \rangle = \langle p^s Vv, (F - V)v \rangle \in W(R).$$
(7.10.4)

Then similarly to Case 1 we again get B). Q.E.D.

7.11. Direct calculation of the truncation morphisms.

Lemma. *Suppose* $\mathfrak{U}_m$ *has a fine moduli scheme* $\mathcal{U}_m$. *Given an object* $\{M'_{g-3} \supset \dots \supset M'_0; M_{g-1} \supset \dots \supset M_m\}$ *of* $\mathfrak{U}_m$ *over* $S = \mathrm{Spec}(R)$, *to extend it to an object of* $\mathfrak{U}_{m-1}$ *over* S *is equivalent to giving elements* $t, x_1, \dots, x_{m-1}, x \in R$ *satisfying*

$$x_i^{p^2} - x_i - a_i t - b_i = 0 \ (1 \le i \le m - 1), \quad f = 0$$
(7.11.1)

where

$$f = \begin{cases} x^{p^2} - x - \sum_{i<g/2} t(a_{g-2-i}x_i - a_i x_{g-2-i}) & \text{if } m = g - 1 \\ x^p - x - \sum_i (c_i x_i^p - c_i^p x_i) - d & \text{if } m \text{ is even} \\ x^{p^2} - x - \sum_i (c_i x_i^{p^2} - c_i^{p^2} x_i) - d & \text{if } m \text{ is odd, } 1 < m < g - 1 \\ t & \text{if } m = 1 \end{cases}$$
(7.11.2)

when g *is even, and when* g *is odd*

$$f = \begin{cases} x^p - x - \sum_i e x_i^{p+1} & \text{if } m = g - 1 \\ x^p - x - \sum_i e(c_i x_i^p - c_i^p x_i) - d & \text{if } m \text{ is even, } m < g - 1 \\ x^{p^2} - x - \sum_i e(c_i x_i^{p^2} - c_i^{p^2} x_i) - d & \text{if } m \text{ is odd, } m > 1 \\ t & \text{if } m = 1 \end{cases}$$
(7.11.3)

and a_i, b_i, c_i, d, e *are canonical (i.e. pull back of the corresponding sections of* $O_{\mathcal{U}_m}$ *).*

Proof. Again denote $A_R = W(R) \otimes_W H$. As in the proof of Lemma 7.9 we can choose

$$v_m = x + \zeta y + \sum_i \zeta_i x_i \in M_m.$$
(7.11.4)

46

Let $v = v_m + w + \beta F^{g-m-1} y$, where

$$w = \sum_{j \geq g-m} \beta_j F^{g-m-1} x_j. \tag{7.11.5}$$

By condition d) we can choose $\lambda_1, ..., \lambda_{g-m-1} \in A_R$ such that

$$(F - V)v_m (\bmod A_R y) - \sum_{j < g-m} \lambda_j v_j' = \sum_{j \geq g-m} \mu_j F^{g-m} x_j. \tag{7.11.6}$$

Hence A) is equivalent to that

$$(F - V)w + \sum_{j \geq g-m} \mu_j F^{g-m} x_j \equiv \tau v_{g-m}' \quad (\bmod F^{g-m} M_{g-3}') \tag{7.11.7}$$

for some $\tau \in A_R$, or explicitly

$$\bar{\beta}_j^{p^2} - \bar{\beta}_j = \bar{\alpha}_{(g-m)j} \bar{\tau} - \bar{\mu}_j \quad (g - m \leq j \leq g - 2). \tag{7.11.8}$$

And condition B') can be written explicitly as (for $m > 1$ only)

$$\bar{\beta}^{p^{m-1-2s}} - \bar{\beta} = \overline{\langle v_m + w, p^{s+1} F^{m-1-2s}(v_m + w) \rangle}. \tag{7.11.9}$$

When g is even, the right hand side of (7.11.9) is equal to

$$\begin{cases} \displaystyle\sum_{j \geq g/2} (\bar{\beta}_j^{p^2} \bar{\beta}_{g-2-j} - \bar{\beta}_j \bar{\beta}_{g-2-j}^{p^2}) & \text{if } m = g - 1, \\[2ex] \displaystyle\sum_{j \geq g-m} (\bar{\beta}_j^{p} \bar{\zeta}_{g-2-j} - \bar{\beta}_j \bar{\zeta}_{g-2-j}^{p}) + \overline{\langle v_m, p^{s+1} F v_m \rangle} & \text{if } m \text{ is even}, \\[2ex] \displaystyle\sum_{j \geq g-m} (\bar{\beta}_j^{p^2} \bar{\zeta}_{g-2-j} - \bar{\beta}_j \bar{\zeta}_{g-2-j}^{p^2}) + \overline{\langle v_m, p^{s+1} F^2 v_m \rangle} & \text{if } m \text{ is odd}, m < g - 1. \end{cases} \tag{7.11.10}$$

When g is odd, the right hand side of (7.11.9) is equal to

$$\begin{cases} \displaystyle\sum_{j} \bar{\epsilon} \bar{\beta}_j^{p+1} & \text{if } m = g - 1, \\[2ex] \displaystyle\sum_{j \geq g-m} \bar{\epsilon}(\bar{\beta}_j^{p} \bar{\zeta}_j - \bar{\beta}_j \bar{\zeta}_j^{p}) + \overline{\langle v_m, p^{s+1} F v_m \rangle} & \text{if } m \text{ is even}, m < g - 1, \\[2ex] \displaystyle\sum_{j \geq g-m} \bar{\epsilon}(\bar{\beta}_j^{p^2} \bar{\zeta}_j - \bar{\beta}_j \bar{\zeta}_j^{p^2}) + \overline{\langle v_m, p^{s+1} F^2 v_m \rangle} & \text{if } m \text{ is odd}. \end{cases} \tag{7.11.11}$$

Let

$$a_i = \bar{\alpha}_{(g-1)(g-m-1+i)}, \quad b_i = \bar{\mu}_{g-m-1+i} \ (1 \leq i \leq m - 1),$$
$$d = \overline{\langle v_m, p^{s+1} F^{m-1-2s} v_m \rangle}, \quad e = \bar{\epsilon}, \tag{7.11.12}$$

$$c_i = \begin{cases} \bar{\zeta}_{m-i} & \text{if } g \text{ is even} \\ \bar{\zeta}_{g-m-1+i} & \text{if } g \text{ is odd} \end{cases} \quad (1 \leq i \leq m - 1) \tag{7.11.13}$$

47

and

$$t = \bar{\tau}, \; x = \bar{\beta}, \; x_i = \bar{\beta}_{g-m-1+i} \; (1 \le i \le m-1). \tag{7.11.14}$$

(Note that we have $\epsilon \in F_{p^2}$, $\epsilon^p = -\epsilon \ne 0$.) Then (7.11.8) and (7.11.9) show that t, x, x_i satisfy (7.11.1), and $a_i, b_i, c_i, d, \epsilon$ are clearly canonical. Conversely, given t, x, x_i satisfying (7.11.1), we lift them to $\tau, \beta, \beta_{g-m-1+i} \in W(R)$ respectively. Then the above argument shows that $v = v_m + w + \beta F^{g-m-1} y$ gives an object of $\mathfrak{U}_{m-1}$ extending $\{M'_{g-3} \supset \dots \supset M'_0; M_{g-1} \supset \dots \supset M_m\}$. This gives a one to one correspondence by Lemma 7.9 and Lemma 7.10.   Q.E.D.

7.12. Smoothness of the truncation morphisms.

Corollary. *If $\mathfrak{U}_m$ has a fine moduli scheme $\mathcal{U}_m$, Then $\mathfrak{U}_{m-1}$ has a fine moduli scheme $\mathcal{U}_{m-1}$. Furthermore, the canonical morphism $\mathcal{U}_{m-1} \to \mathcal{U}_m$ induced by t_m is a smooth surjective morphism of relative dimension 1.*

Proof. By t_m, every object of $\mathfrak{U}_{m-1}$ over S gives a morphism $S \to \mathcal{U}_m$. By Lemma 7.11, the functor

$((\mathcal{U}_m\text{-schemes})) \to ((\text{sets}))$

$(f : S \to \mathcal{U}_m) \mapsto \{\text{objects in } \mathfrak{U}_{m-1} \text{ extending the object of } \mathfrak{U}_m$
$\qquad\qquad\qquad\qquad \text{represented by } f\}$

is represented by a $\mathcal{U}_m$-scheme $\mathcal{U}_{m-1}$. Locally, the morphism $\mathcal{U}_{m-1} \to \mathcal{U}_m$ induced by t_m is given by variables t, x, x_i satisfying the defining relations (7.11.1), hence is clearly a smooth surjective morphism of relative dimension 1.   Q.E.D.

7.13. Smoothness of the moduli schemes.

Corollary. *Each $\mathfrak{U}_m$ has a fine moduli scheme $\mathcal{U}_m$, and $\mathfrak{N}_g$ has a fine moduli scheme which is smooth of dimension $[g^2/4]$ over F_{p^4}.*

Proof. Clearly $\mathcal{N}_0$ and $\mathcal{N}_1$ exist and are isomorphic to $\text{Spec}(F_{p^4})$. Hence each $\mathcal{U}_m$ exists by double induction and Corollary 7.12 (recall that $\mathcal{U}_{g-1}$ is an open subscheme of $\mathcal{N}_{g-2}$). Therefore $\mathfrak{N}_g$ has a fine moduli scheme $\mathcal{N}_g$ and $\mathcal{U}_0^\Theta$'s form an open cover of $\mathcal{U}_{g-1}$ (see 7.6 above). Furthermore, by induction we may assume $\mathcal{N}_{g-2}$ is smooth over F_{p^4} of dimension $[(g-2)^2/4]$. Since each canonical morphism $\mathcal{U}_{m-1} \to \mathcal{U}_m$ is smooth of relative dimension 1 by Corollary 7.12, we see $\mathcal{N}_g$ is smooth over F_{p^4} of dimension

$$[\frac{(g-2)^2}{4}] + (g-1) = [\frac{g^2}{4}]. \tag{7.13.1}$$

Q.E.D.

7.14. The moduli of PFTQs of Dieudonné modules is integral.

Lemma. *For $g \ge 1$, the scheme $\mathcal{N}_g$ is geometrically integral. Furthermore, if $\{M_{g-1} \supset \dots \supset M_0\}$ is the universal PFTQ of Dieudonné modules over $\mathcal{N}_g$, then the generic fiber of M_0 over $\mathcal{N}_g$ is supergeneral.*

48

Proof. Again we use double induction. First we assume $\mathcal{N}_{g-2}$ is geometrically integral, and for the universal PFTQ of Dieudonné modules $\{M'_{g-3} \supset ... \supset M'_0\}$ over $\mathcal{N}_{g-2}$, the generic fiber of M'_0 over $\mathcal{N}_{g-2}$ is supergeneral.

Note that in the proof of Lemma 7.11, when $m = g - 1$ we have

$$x_i^{p^2} - x_i = a_i t \quad \text{and} \quad c_i = x_{g-2-i} \ (\forall i), \tag{7.14.1}$$

where t is transcendental over $K_{g-1} := K(\mathcal{N}_{g-2})$, (the function field of $\mathcal{N}_{g-2}$). By Fact 5.8, the generic fiber of M'_0 is supergeneral iff $\bar{\alpha}_{1j}$ $(1 \le j \le g - 2)$ (see (7.6.5)) are linearly independent over $\mathbf{F}_{p^2}$. This shows that $\bar{\alpha}_{ij}$ $(i \le j \le g - 2)$ (see (7.6.5)) are linearly independent over $\mathbf{F}_{p^2}$, also by Fact 5.8.

Suppose $\mathcal{U}_m$ is integral (by induction). Let $K_m = K(\mathcal{U}_m)$. By Corollary 7.8,

$$K' = K_m(t)[x_1, ..., x_{m-1}]/(x_i^{p^2} - x_i - a_i t - b_i | \forall i) \tag{7.14.2}$$

is a domain. In particular, when $m = g - 1$, this shows by Corollary 7.8 that $c_1, ..., c_{g-2}, c_1^p, ..., c_{g-2}^p$ are linearly independent over K_{g-1}.

We now show that f is irreducible over K'. Let $K'' = K_m(t)$. Then $K' = K''(\theta_1, ..., \theta_{m-1})$, where $\theta_i^{p^2} - \theta_i = a_i t + b_i$. We only do the case when g is even and $m < g - 1$. The arguments for the cases when $m = g - 1$ and that when g is odd are similar, and are left to the reader.

Case 1: m is even. We need only show that f has no zero in K'. Suppose $\theta \in K'$ such that $f(\theta) = 0$. Take $\phi \in \text{Gal}(K'/K''(\theta_1, ..., \theta_{m-2}))$ such that $\phi(\theta_{m-1}) = \theta_{m-1} + 1$. Then $\theta' = \phi(\theta) - \theta$ satisfies $\theta'^p - \theta' = c_{m-1} - c_{m-1}^p$. Hence $\theta' = \epsilon - c_{m-1}$ for some $\epsilon \in \mathbf{F}_{p^2}$. Now $\theta - (\epsilon - c_{m-1})\theta_{m-1}$ is invariant under ϕ, hence is contained in an intermediate field between K' and $K''(\theta_1, ..., \theta_{m-2})$. Repeating this argument we get

$$\theta = (\epsilon_{m-1} - c_{m-1})\theta_{m-1}^p + \epsilon_{m-1}^p \theta_{m-1} + \theta', \tag{7.14.3}$$

where $\epsilon_{m-1} \in \mathbf{F}_{p^2}$ and $\theta' \in K''(\theta_1, ..., \theta_{m-2})$. By induction we get

$$\theta = \sum_i ((\epsilon_i - c_i)\theta_i^p + \epsilon_i^p \theta_i) + s, \tag{7.14.4}$$

where $\epsilon_i \in \mathbf{F}_{p^2}$ and $s \in K''$. Thus

$$f(\theta) = \sum_i (\epsilon_i^p - c_i^p)(a_i t + b_i) + s^p - s - d = 0. \tag{7.14.5}$$

Since $c_1^p, ..., c_{m-1}^p$ are linearly independent over K_{g-1} (see above), we have $\sum_i a_i c_i^p \ne 0$ and (7.14.5) is impossible.

Case 2: m is odd and > 1. Then f is irreducible iff $x^p - x - \epsilon(\sum_i (c_i x_i^{p^2} - c_i^{p^2} x_i) - d)$ has no zero in K' for any $\epsilon \in \mathbf{F}_{p^2}^*$. Substitute $\theta_i^{p^2} - \theta_i = a_i t + b_i$ to f. The remaining arguments are similar to Case 1.

Finally we show that the generic fiber of M_0 over $\mathcal{N}_g$ is supergeneral, or equivalently, in the case when $m = g - 1$, the above $x_1, ..., x_{g-2}, x$ and 1 are linearly independent over $\mathbf{F}_{p^2}$. Suppose we have a linear equation over $\mathbf{F}_{p^2}$:

$$l_1 x_1 + ... + l_{g-2} x_{g-2} + l_{g-1} x + l_g = 0 \tag{7.14.6}$$

Take the p^2th power of (7.14.6) minus (7.14.6), we get a linear equation over $\mathbb{F}_{p^2}$ of $a_1 t, \ldots, a_{g-2} t$ and $\sum_{i<g/2} t(a_{g-2-i} x_i - a_i x_{g-2-i})$ when g is even (resp. $t \sum_i a_i x_i^p$ when g is odd). This shows that $l_1 = \ldots = l_g = 0$. Q.E.D.

7.15. Proof of Proposition 7.6.

The smoothness and dimension of $\mathcal{N}_g$ are given in Corollary 7.13. The fact that $\mathcal{N}_g$ is integral and the last statement of Proposition 7.6 are given by Lemma 7.14. Q.E.D.

7.16. The weak form of Proposition 4.3.

Remark. We have now proved the weak form of Proposition 4.3 (see 4.11). Indeed. by 7.1 and 3.10 we have a canonical morphism $\mathcal{P}'_g \to \mathcal{N}_g$, and by Remark 7.5 we see that $F^{g-1}_{\mathcal{N}_g/\mathbb{F}_{p^4}}$ factors through $\mathcal{P}'_g$ (i.e. $\mathcal{P}'_g$ and $\mathcal{N}_g$ differ by at most a purely inseparable extension). Hence $\mathcal{P}'_g$ is geometrically irreducible of the same dimension as that of $\mathcal{N}_g$, i.e. $[g^2/4]$.

8. Some class numbers

In this chapter we study class numbers $H_g(p, 1)$, $H_g(1, p)$, their generalizations, and then prove Proposition 4.7 and Corollary 4.8.

8.1. Hermitian forms over the quaternion algebra.

We use the same notation as in 4.6. One can take an imaginary quadratic extension $K = \mathbb{Q}[\alpha]$, unramified at p, such that

$$B \cong K[F]/(F^2 + p, F\alpha - \alpha^\sigma F), \tag{8.1.1}$$

where $\sigma \in \mathrm{Gal}(K/\mathbb{Q})$ is the complex conjugation (we reserve bar ($^-$) for the involution of B). Furthermore, we have

$$K_p = K \otimes \mathbb{Z}_p \cong W(\mathbb{F}_{p^2}) \otimes \mathbb{Q}. \tag{8.1.2}$$

For any $\beta = a + bF \in B$ ($a, b \in K$), denote by $|\beta| = \beta\bar\beta = aa^\sigma + pbb^\sigma$, the absolute value of β.

Let $\mathcal{O}$ be a maximal order in B and $H = \mathcal{O} \otimes \mathbb{Z}_p$, as in 5.7. Then (5.7.1) holds, in which σ is as above, and σ is also the lifting of the absolute Frobenius of $\mathbb{F}_{p^2}$. Note that $|\ |$ gives a surjective map $H \to \mathbb{Z}_p$ (see [45, Lemma 1.2]).

Let $\langle , \rangle$ be a hermitian form on $N = B^{\oplus g}$. For $u, v \in N$, write $\langle u, v \rangle = R(u, v) + J(u, v)F$, where $R(u, v), J(u, v) \in K$. Then we have

$$R(u, v) + J(u, v)F = \langle u, v \rangle = \overline{\langle v, u \rangle} = R(v, u)^\sigma - J(v, u)F \tag{8.1.3}$$

and

$$\begin{aligned}
R(Fu, v) + J(Fu, v)F &= \langle Fu, v \rangle = F\langle u, v \rangle \\
&= F(R(u, v) + J(u, v)F) = R(u, v)^\sigma F - pJ(u, v)^\sigma.
\end{aligned} \tag{8.1.4}$$

Comparing (8.1.3) and (8.1.4) we get

$$J(v, u) = -J(u, v), \quad J(Fu, v) = J(Fv, u)^\sigma \tag{8.1.5}$$

and

$$R(u, v) = R(v, u)^\sigma = J(Fv, u), \quad R(u. Fv) = -pJ(v, u) \tag{8.1.6}$$

Therefore $\langle , \rangle$ is uniquely determined by J. Conversely, if J is a K-bilinear form on N satisfying (8.1.5), then $\langle u, v \rangle = J(Fv, u) + J(u, v)F$ defines a hermitian form $\langle , \rangle$ on N. Thus we get:

Lemma. *There is a canonical one to one correspondence ζ between the set of hermitian forms on N and the set of K-bilinear forms J on N satisfying (8.1.5).*

51

8.2. Equivalence classes of positive definite hermitian forms.

Next we consider a positive definite hermitian form $\langle . \rangle$ on an $\mathcal{O}$-lattice $L \subset N$ such that $\langle L, L \rangle \subset \mathcal{O}$, which corresponds to an element of Σ in (4.6.2).

For any $f, f' \in \Sigma$. it is easy to see that $f \approx f'$ iff $\zeta(f)$ and $\zeta(f')$ are equivalent, i.e. there exists $\gamma \in GL_g(\mathcal{O})$ such that $\zeta(f)(\gamma u, \gamma v) = \zeta(f')(u, v)$ for all $u, v \in L$. Hence we have:

Corollary. *The map ζ in Lemma 8.1 induces a monomorphism from $\Sigma / \approx$ to the set of equivalence classes of K-bilinear forms J on N satisfying (8.1.5).*

8.3. Classification of the p-adic hermitian forms.

We first give a "local" classification of the elements in Σ. i.e. a classification of $T \otimes \mathbf{Z}_l$ (where T is a hermitian form on L, and l is a prime number) up to equivalence. For $l = p$ we can apply Lemma 6.1 to J and get:

Proposition. *Any $f \in \Sigma$ is equivalent to a canonical form $\mathrm{diag}(A_1, A_2, ...)$ in $M_g(\mathcal{O}_p)$, where each A_i is a matrix either equal to (p^r) for some r or equal to*

$$ p^r \begin{pmatrix} 0 & F \\ -F & 0 \end{pmatrix} $$

for some r. Furthermore, this canonical form is unique up to permutations of $A_1, A_2, ...$.

This result is well known in the cases when $\det(f) = 1$ and when $\det(f) = p^{4[(g+1)/2]}$. $p | f^2$ (cf. [12, Proposition 5.1] and [31, Corollary 2.9]).

8.4. Classification of the l-adic hermitian forms.

Using the same method as that in 8.3, we can also show that:

Proposition. *For $l \neq p$. any $f \in \Sigma$ is equivalent to a canonical form $\mathrm{diag}(l^{r_1}, l^{r_2}, ..., l^{r_g})$ for unique $r_1 \leq r_2 \leq ... \leq r_g$ in $M_g(\mathcal{O}_l)$.*

This fact is also partially known (cf. [31, Corollary 2.9]).

8.5. Adelic hermitian forms and class numbers.

By an *adelic hermitian form* we will mean a family

$$ \mathfrak{f} = \{ f_l \in M_g(\mathcal{O}_l) | f_l = f_l^t, l \text{ prime} \} \tag{8.5.1} $$

such that $f_l = I_g$ for almost all l. Denote by

$$ \mathcal{L}_{\mathfrak{f}, \sim} = \{ f \in \Sigma | f \sim f_l \text{ in } M_g(\mathcal{O}_l) \ \forall l \}, \tag{8.5.2} $$

and
$$\mathcal{L}_{\mathfrak{f},\approx} = \{f \in \Sigma | f \approx f_l \text{ in } M_g(\mathcal{O}_l) \; \forall l\}. \tag{8.5.3}$$

By Proposition 8.3 and Proposition 8.4, for $\mathcal{L}_{\mathfrak{f},\sim}$ and $\mathcal{L}_{\mathfrak{f},\approx}$ we can take each f_l to be a canonical form. Furthermore, note that $l|f$ in $M_g(\mathcal{O})$ iff $l|f$ in $M_g(\mathcal{O}_l)$, hence for $\mathcal{L}_{\mathfrak{f},\sim}$ we can take $\mathfrak{f}$ such that $l \nmid f_l$ for each l, such an adelic form is called *reduced*.

Suppose $f, f' \in \mathcal{L}_{\mathfrak{f},\approx}$. If $f \sim f'$, then $f \approx f'$ because $\det(f) = \det(f')$. Hence we have:

Proposition. *The inclusion* $\mathcal{L}_{\mathfrak{f},\approx} \hookrightarrow \mathcal{L}_{\mathfrak{f},\sim}$ *induces a one to one correspondence* $\mathcal{L}_{\mathfrak{f},\approx}/\approx \; \to \; \mathcal{L}_{\mathfrak{f},\sim}/\sim$. *Furthermore, if* $\mathfrak{f}'$ *is the reduced adelic form quasi-equivalent to* $\mathfrak{f}$, *then* $\#(\mathcal{L}_{\mathfrak{f},\approx}/\approx) = \#(\mathcal{L}_{\mathfrak{f}',\approx}/\approx)$.

Number theorists already have some results on $\mathcal{L}_{\mathfrak{f},\sim}/\sim$ (cf. [24, p. 490], [25, p. 393]), while in the study of moduli spaces we need to use $\mathcal{L}_{\mathfrak{f},\approx}/\approx$. The above Proposition allows us to make use of the results on quasi-equivalent classes of hermitian forms for our purpose, as we will see in the following.

Remark. Any $f \in \Sigma$ can be explained as an $\mathcal{O}$-semilinear monomorphism $\lambda_f : L \to L^\vee$, where $L^\vee$ is the dual lattice of L. Denote $\Gamma_f = L^\vee/\lambda_f(L)$, which is a finite abelian group. Clearly $\Gamma_f \cong \Gamma_{f'}$ if $f \approx f'$. By Lemma 8.1 we see that there is an induced non-degenerate alternating form on Γ_f. In particular $\det(f) = |\Gamma_f| = d^4$ for some d.

8.6. Duality of the class numbers.

It is also easy to prove the following:

Proposition ("duality"). *For any* $\mathfrak{f}$ *we have* $\#(\mathcal{L}_{\mathfrak{f},\sim}/\sim) = \#(\mathcal{L}_{\mathfrak{f}^{-1},\sim}/\sim)$.

8.7. Polarizations of superspecial abelian varieties.

Next we explain the connection between certain hermitian forms and polarizations of superspecial abelian varieties.

Again let E be a supersingular elliptic curve as in 1.2. Fix a principal polarization μ_0 of E and let $\bar{f} = \mu_0^{-1} \circ f^t \circ \mu_0$ for any $f \in \text{End}(E \otimes \bar{\mathbb{F}}_p)$. This gives an identification $\text{End}^0(E \otimes \bar{\mathbb{F}}_p) \cong B$ in which F plays the role as in (8.1.1). Hence $\text{End}(E^g \otimes \bar{\mathbb{F}}_p) \cong M_g(\mathcal{O})$. Furthermore, $\mu = \mu_0^g$ is a principal polarization of $E^g \otimes \bar{\mathbb{F}}_p$.

Let Π be the set of polarizations of $E^g \otimes \bar{\mathbb{F}}_p$. By [31, Proposition 2.8], there is a one-to-one correspondence
$$\Phi : \Pi \xrightarrow{\sim} \Sigma \subset \text{End}(E^g \otimes \bar{\mathbb{F}}_p)$$
$$\eta \mapsto \mu^{-1} \circ \eta$$

which induces a one to one correspondence of equivalence classes.

Let $\eta \in \Pi$ and $G = \ker(\eta)$. As in 3.7, η induces an isomorphism
$$\theta_\eta : G \xrightarrow{\sim} \ker(\eta^t) \cong G^D \tag{8.7.1}$$

which is called a *quasi-polarization* of G.

Lemma. *For any adelic form $\mathfrak{f}$ and any $\eta, \eta' \in \Pi$ such that*

$$\Phi(\eta), \Phi(\eta') \in \mathcal{L}_{\mathfrak{f}, \approx}, \tag{8.7.2}$$

there is an isomorphism

$$\phi : G' = \ker(\eta') \to G = \ker(\eta) \tag{8.7.3}$$

such that θ_η and $\theta_{\eta'}$ are equivalent via ϕ, i.e. $\theta_{\eta'} = \phi^D \circ \theta_\eta \circ \phi$.

Proof. For each prime number l, take $\psi_l \in GL_g(\mathcal{O}_l)$ such that $\Phi(\eta') = \psi_l^t \circ \Phi(\eta) \circ \psi_l$ in $M_g(\mathcal{O}_l)$. Let $d = \deg(\eta)$. We can take $\psi \in \text{End}(E^g \otimes \bar{\mathbb{F}}_p) \cong M_g(\mathcal{O})$ such that $l^r | (\psi - \psi_l)$ in $M_g(\mathcal{O}_l)$ for any l if $l^r | d$. Hence ψ induces an automorphism of $(E^g \otimes \bar{\mathbb{F}}_p)[d]$ and $\eta' \equiv \psi^t \circ \eta \circ \psi \pmod{d}$. This shows $G = \psi(G')$. Therefore ψ induces an isomorphism $\phi : G' \to G$ such that $\theta_{\eta'} = \phi^D \circ \theta_\eta \circ \phi$. Q.E.D.

In other words, the pair (G, θ_η) is uniquely determined by $\mathfrak{f}$. Conversely, suppose $\eta \in \Pi$ and $\mathfrak{f}$ is the adelic form of η (i.e. $\mathfrak{f}_l \approx \Phi(\eta)$ in $M_g(\mathcal{O}_l)$ for all l), then from Proposition 8.3 and Proposition 8.4 we see that $\mathfrak{f}$ is uniquely determined by $G = \ker(\eta)$. Summarizing these we get

Theorem. *Let $\eta \in \Pi$, $G = \ker(\eta)$ and $\mathfrak{f}$ be the adelic form of η. Then Φ induces a one to one correspondence between $\Pi_G := \{\eta' \in \Pi | \ker(\eta') \cong G\}$ and $\mathcal{L}_{\mathfrak{f}, \approx}$. Hence*

$$\#(\Pi_G/\sim) = \#(\mathcal{L}_{\mathfrak{f}, \approx}/\approx) = \#(\mathcal{L}_{\mathfrak{f}, \sim}/\sim). \tag{8.7.4}$$

Furthermore, G can be decomposed to a direct product of some $(E \otimes \bar{\mathbb{F}}_p)[l^r]$'s and some $(E^2 \otimes \bar{\mathbb{F}}_p)[F^s]$'s ($s$ odd) such that θ_η is equal to the product of quasi-polarizations of the direct factors. In particular, if $\deg(\eta) = d^2$, then $\det(\Phi(\eta)) = d^4$.

Remark. If we fix an isomorphism $\text{End}^0(E \otimes \bar{\mathbb{F}}_p) \cong B$ and a μ_0, then we have a "standard" one to one correspondence in Theorem 8.7.

8.8. Proof of Proposition 4.7.

Let $A = \begin{pmatrix} 0 & F \\ -F & 0 \end{pmatrix}$, $B = (p)$ when g is odd and $B = A$ when g is even. Let $\mathfrak{f} = \{\mathfrak{f}_l \in M_g(\mathcal{O}_l)\}$, where $\mathfrak{f}_l = I_g$ for all $l \neq p$ and $\mathfrak{f}_p = \text{diag}(A, ..., A, B)$. Then by definition $\mathcal{L}_{\mathfrak{f}, \sim} = \mathcal{L}_g(1, p)$.

If $\eta \in \Pi$ such that $\ker(F) \subset \ker(\eta)$ and $\deg(\eta) = p^{2r}$ ($r = [(g+1)/2]$), then by Theorem 8.7 it is easy to see that $\Phi(\eta) \in \mathcal{L}_{\mathfrak{f}, \approx}$ (in the case g=2 this was already shown in [31, Theorem 2.1]). Hence Proposition 4.7 is a special case of (8.7.4). Q.E.D.

8.9. Proof of Corollary 4.8.

For i) and ii), apply Proposition 8.5 to Fact 4.6 and Proposition 4.7 respectively. For iii), apply Proposition 8.5 and Proposition 8.6 to Proposition 4.7. Q.E.D.

9. Examples on $\mathcal{S}_{g,1}$

In this chapter we describe $\mathcal{P}_g$ (or $\mathcal{P}_{g,\eta}$) and $\mathcal{P}'_g$ for some low values of g and any characteristic p, and use the results to study the structure of the locus $\mathcal{S}_{g,1}$ of principally polarized abelian varieties of dimension g.

9.1. Example, $g = 1$.

When $g = 1$, the set

$$\mathcal{S}_{1,1}(\bar{\mathbf{F}}_p) \subset \mathcal{A}_{1,1} \otimes \bar{\mathbf{F}}_p \cong \mathbf{A}^1 \tag{9.1.1}$$

is the set of supersingular j-invariants. We write

$$h_p := \#(\mathcal{S}_{1,1}(\bar{\mathbf{F}}_p)). \tag{9.1.2}$$

This number equals the class number of $B = Q_{\infty,p}$ (see (1.2.5)), which is equal to

$$h_p = \frac{p-1}{12} + \{1 - (\frac{-3}{p})\}/3 + \{1 - (\frac{-4}{p})\}/4 \tag{9.1.3}$$

(cf. [9, p. 200] and [29, p. 312]), as was proved by Deuring (using a class number computation by Eichler), and later proved along different lines by Igusa, see [29, p. 312]. Explicitly: $h_2 = h_3 = 1$ and for $p \geq 5$,

$$h_p = [\frac{p-1}{12}] + \begin{cases} 0 & p \equiv 1 \ (\mathrm{mod}\ 12), \\ 1 & p \equiv 5 \ \mathrm{or}\ 7 \ (\mathrm{mod}\ 12), \\ 2 & p \equiv 11 \ (\mathrm{mod}\ 12). \end{cases} \tag{9.1.4}$$

This can also be expressed by the mass formula:

$$\sum \frac{1}{\#(\mathrm{Aut}(C))} = \frac{p-1}{24}, \tag{9.1.5}$$

where the summation is over all isomorphism classes of supersingular elliptic curves C over $\bar{\mathbf{F}}_p$.

9.2. Example, $g = 2$.

For $g = 2$, an FTQ over k is of the form

$$\rho_1 : E^2 \otimes k \cong Y_1 \to Y_0, \quad \ker(\rho_1) \cong \alpha_p. \tag{9.2.1}$$

Such an FTQ is automatically rigid. For any η satisfying (3.6.1) (i.e. $\ker(\eta) = E^2[F] \otimes k$), (9.2.1) is automatically a PFTQ with respect to η, hence

$$\mathcal{P}_{2,\eta} \cong \mathcal{P}'_{2,\eta} \cong \mathbf{P}^1 \tag{9.2.2}$$

(see Example 3.8). The number of irreducible components of $S_{2,1} \otimes k$ is equal to $H_2(1,p)$ (see [35, Theorem 5.7]). This number was explicitly calculated by Hashimoto and Ibukiyama (see [25, p.696]). It is equal to 1 when $p = 2, 3$ or 5, and when $p > 5$,

$$
H_2(1,p) = (p^2 - 1)/2880 + (p+1)\left(1 - \left(\frac{-1}{p}\right)\right)/64
$$

$$
+ 5(p-1)\left(1 + \left(\frac{-1}{p}\right)\right)/192 + (p+1)\left(1 - \left(\frac{-3}{p}\right)\right)/72 \qquad (9.2.3)
$$

$$
+ (p-1)\left(1 + \left(\frac{-3}{p}\right)\right)/36
$$

$$
+ \begin{cases} 2/5 & \text{if } p \equiv 2 \text{ or } 3 \pmod 5 \\ 0 & \text{if } p \equiv 1 \text{ or } 4 \pmod 5 \end{cases}
$$

$$
+ \begin{cases} 1/4 & \text{if } p \equiv 3 \text{ or } 5 \pmod 8 \\ 0 & \text{if } p \equiv 1 \text{ or } 7 \pmod 8 \end{cases}
$$

$$
+ \begin{cases} 1/6 & \text{if } p \equiv 5 \pmod{12} \\ 0 & \text{if } p \equiv 1, 7 \text{ or } 11 \pmod{12} \end{cases}
$$

where $\left(\frac{q}{p}\right)$ denotes the Legendre symbol.

Let η be a polarization of $E^2 \otimes k$ such that $\ker(\eta) = E^2[F] \otimes k$. Then $G_\eta = \mathrm{Aut}(E^2 \otimes k, \eta)/\{\pm 1\}$ is isomorphic to one of the following groups:

$$
\{1\},\ \mathbf{Z}/2\mathbf{Z},\ \mathbf{Z}/3\mathbf{Z},\ V_4 \cong \mathbf{Z}/2\mathbf{Z} \times \mathbf{Z}/2\mathbf{Z},\ S_3,\ A_4,\ D_{12},\ S_4,\ A_5. \qquad (9.2.4)
$$

Let $W_\eta \subset S_{2,1}$ be the irreducible component corresponding to η (i.e. the closure of $\Psi(\mathcal{P}'_{2,\eta})$, see (4.2.1)). Then the action of G_η on $\mathcal{P}_{2,\eta}$ is generically free, and we have

$$
\mathbf{P}^1 \cong \mathcal{P}_{2,\eta} \to \mathcal{P}_{2,\eta}/G_\eta \cong \tilde{W}_\eta \to W_\eta, \qquad (9.2.5)
$$

where $\tilde{W}_\eta$ is the normalization of W_η (cf. [35, Section 7, 8.1]). By [32, Theorem 7.1] we see that those in (9.2.4) are exactly the groups which do appear in this way.

Conclusion. Let Λ be a set of representatives of equivalence classes of polarizations η of $E^2 \otimes \bar{\mathbf{F}}_p$ satisfying $\ker(\eta) = E^2[F] \otimes \bar{\mathbf{F}}_p$. Then there is a one to one correspondence ψ between Λ and the set of irreducible components of $S_{2,1} \otimes \bar{\mathbf{F}}_p$. Denote by W_η the irreducible component corresponding to η under ψ. The normalization of W_η is isomorphic to $\mathcal{P}_{2,\eta}/G_\eta$, where $\mathcal{P}_{2,\eta} \cong \mathbf{P}^1$ and $G_\eta = \mathrm{Aut}(E^2 \otimes \bar{\mathbf{F}}_p, \eta)/\{\pm 1\}$. We have $\#(\Lambda) = H_2(1,p)$ and

$$
S_{2,1} \otimes \bar{\mathbf{F}}_p = \bigcup_{\eta \in \Lambda} W_\eta. \qquad (9.2.6)
$$

9.3. Calculation via the truncation morphisms.

When $g > 2$, we proceed as follows. Let $\mathcal{V}_m$ be the fine moduli scheme of the category $\mathfrak{V}_m$ of truncated PFTQs $\{S; Y_i, \eta_i(m \le i < g); \rho_i(m < i < g)\}$. (This moduli scheme exists by the same argument as that in Lemma 3.7.) Then we can calculate $\mathcal{V}_m$'s inductively. First we note the following two facts:

i) $\mathcal{V}_{g-2}$ is easy to calculate: To give a Y_{g-2} from $Y_{g-1} = E^g \times S$ is equivalent to giving a flat subgroup scheme $G \subset \alpha_p^g \times S$ of α-rank $g - 1$ such that condition ii) in Definition 3.9 holds. This is then equivalent to choosing a section $(x_1, ..., x_g)$ of the α-sheaf of $\alpha_p \times S$ such that the following $[(g-1)/2]$ equations are satisfied:

$$\sum_i x_i^{p^{g-2j}+1} = 0 \quad (0 < j < g/2) \tag{9.3.1}$$

when g is odd, and

$$\sum_{i \le g/2} (x_i x_{g-i}^{p^{g-2j}} - x_{g-i} x_i^{p^{g-2j}}) = 0 \quad (0 < j < g/2) \tag{9.3.2}$$

when g is even.

ii) It is also easy to determine $\mathcal{V}_0 = \mathcal{P}_{g,\eta}$ from $\mathcal{V}_1$: Since $G = \ker(Y_1 \to Y_1^t)$ is a self-dual α-group of α-rank 2, every flat subgroup scheme of G of α-rank 1 is isotropic. Hence to give a Y_0 is equivalent to giving a flat quotient of rank 1 of the α-sheaf of G. Therefore $\mathcal{V}_0$ is a $\mathbf{P}^1$-bundle over $\mathcal{V}_1$.

Remark. From (9.3.1) and (9.3.2) we see that $\mathcal{V}_{g-2}$ is singular (at a point where all $x_i \in \mathbf{F}_{p^2}$) when $g \ge 5$. Hence there is in general no hope to prove the smoothness of $\mathcal{P}'_g$ over $\mathbf{F}_{p^2}$ using the factorization $\mathcal{V}_0 \to ... \to \mathcal{V}_{g-1}$. Therefore we will use another factorization to prove Proposition 4.3.i) (see 11.3 and 9.7).

By the proof of Lemma 7.11, the truncation morphism $\mathcal{P}'_g \to \mathcal{V}_{g-2}$ is an epimorphism. Hence we have:

Proposition. *The subscheme $T_g \subset \mathbf{P}^{g-1}$ defined by the homogeneous equations in (9.3.1) (when g is odd) or (9.3.2) (when g is even) is irreducible of dimension $[g/2]$. Furthermore, a geometric point $(a_1, ..., a_g) \in T_g$ is non-singular iff the $\mathbf{F}_{p^2}$-linear space generated by $a_1, ..., a_g$ has dimension $\ge [(g-1)/2]$ over $\mathbf{F}_{p^2}$.*

For the second statement, by taking differentials, it reduces to an application of Fact 5.8.

9.4. Example, $g = 3$.

Let

$$E^3 \otimes k = Y_2 \xrightarrow{\rho_2} Y_1 \to Y_0 \tag{9.4.1}$$

be a PFTQ with respect to η, where η satisfies

$$\ker(\eta : E^3 \otimes k \to (E^3 \otimes k)^t) = E^3[p] \otimes k. \tag{9.4.2}$$

Note that

$$(\alpha_p^2 \cong \ker(\rho_2) \subset E^3[F]) \in \mathrm{Grass}_{2,3} \cong \mathbf{P}^2 \qquad (9.4.3)$$

and that $\mathfrak{V}_1$ is represented by the Fermat curve:

$$\rho_2 \in \mathcal{V}_1 = \mathcal{Z}(X^{p+1} + Y^{p+1} + Z^{p+1}) \subset \mathbf{P}^2 \qquad (9.4.4)$$

(see (9.3.2)) and a flat subgroup scheme $H \subset \alpha_p^3 \times \mathcal{V}_1$. The α-sheaf of $H_1 = \alpha_p^3 \times \mathcal{V}_1/H$ is isomorphic to the subsheaf of $O_{\mathcal{V}_1}^{\oplus 3}$ consisting of sections (a, b, c) such that $(a : b : c) = (X : Y : Z)$, hence it is isomorphic to $O_{\mathcal{V}_1}(-1)$.

Let

$$G = \ker(Y_1 \to Y_1^t) = \ker(E^3 \times \mathcal{V}_1/H \to (E^3 \times \mathcal{V}_1/H)^t). \qquad (9.4.5)$$

Then G is an α-group of α-rank 2. Note that η induces an isomorphism $G \cong G^t$. Hence we have $G/H_1 \cong H_1^t$, whose α-sheaf is therefore isomorphic to $O_{\mathcal{V}_1}(1)$.

Let $\mathcal{F}$ be the α-sheaf of G. Then $\mathcal{F}$ is an extension of $O_{\mathcal{V}_1}(-1)$ by $O_{\mathcal{V}_1}(1)$. Since the structure sheaf $\mathcal{E}$ of $\ker(\eta) \times \mathcal{V}_1$ is trivial, the α-sheaf of $\alpha_p^3 \times \mathcal{V}_1$ can be lifted to a subsheaf of $\mathcal{E}$. Hence the α-sheaf of H_1, identified as a subsheaf of the α-sheaf of $\alpha_p^3 \times \mathcal{V}_1$, can also be lifted to a subsheaf of $\mathcal{E}$. Since $\mathcal{F} \cong \omega_{G/\mathcal{V}_1}$, we see that $\mathcal{F} \to O_{\mathcal{V}_1}(-1)$ has a section and hence

$$\mathcal{F} \cong O_{\mathcal{V}_1}(-1) \oplus O_{\mathcal{V}_1}(1). \qquad (9.4.6)$$

By 9.3.ii), $\mathcal{P}_{3,\eta}$ is isomorphic to

$$\mathbf{P}_{\mathcal{V}_1}(O_{\mathcal{V}_1}(-1) \oplus O_{\mathcal{V}_1}(1)) \cong \mathbf{P}_{\mathcal{V}_1}(O_{\mathcal{V}_1} \oplus O_{\mathcal{V}_1}(2)). \qquad (9.4.7)$$

This is a non-singular surface. Thus we have a $\mathbf{P}^1$-fibration

$$\mathcal{P}_{3,\eta} \xrightarrow{\pi} \mathcal{V}_1. \qquad (9.4.8)$$

As in [73, Proposition 2.3], we see that there is a section of π

$$\mathcal{P}_{3,\eta} \supset T \xleftarrow[t]{\sim} \mathcal{V}_1 \qquad (9.4.9)$$

given by

$$t(\rho_2) = (E^3 \otimes k \xrightarrow{\rho_2} Y_1 \to (E^3/E^3[F]) \otimes k = Y_0). \qquad (9.4.10)$$

We have

$$\mathcal{P}'_{3,\eta} = \mathcal{P}_{3,\eta} - T. \qquad (9.4.11)$$

Furthermore, if $x \in \mathcal{P}_{3,\eta}$ represents $\{Y_2 \to Y_1 \to Y_0\}$, then

$$x \in T \implies a(Y_0) = 3, \qquad (9.4.12)$$

$$\pi(x) \in \mathcal{V}_1(\mathbf{F}_{p^2}) \iff a(Y_0) \geq 2, \qquad (9.4.13)$$

$$x \notin T, \pi(x) \notin \mathcal{V}_1(\mathbf{F}_{p^2}) \iff a(Y_0) = 1. \qquad (9.4.14)$$

Remark. The statement (9.4.12) is correct, while in [73, Proposition 2.3] there is a misprint.

Under the morphism
$$\mathcal{P}_{3,\eta} \xrightarrow{\Psi} W_\eta \subset \mathcal{S}_{3,1} \otimes k \qquad (9.4.15)$$
the curve $T \subset \mathcal{P}_{3,\eta}$ is contracted to the point
$$\Psi(T) = (E^3 \otimes k, \eta/p) \in \mathcal{S}_{3,1} \otimes k, \qquad (9.4.16)$$
where η/p is the principal polarization of $(E^3/E^3[F]) \otimes k \cong E^3 \otimes k$ induced by η (as the polarization of Y_0 in (9.4.10)). Outside T the morphism Ψ is finite to one, and generically equals dividing out by the action of $G_\eta = \mathrm{Aut}(E^3 \otimes k, \eta)/\{\pm 1\}$ on $\mathcal{P}_{3,\eta}$. Note that $\Psi(T) \in W_\eta$ is a singular point of W_η. In fact, if $W_\eta^{(n)}$ is an irreducible component of $\mathcal{S}_{g,1,n} \otimes k$ and $x = (E^g \otimes k, \eta/p, \alpha) \in W_\eta^{(n)}$ (where α is a level n-structure), then the tangent space of $W_\eta^{(n)}$ at x has dimension 6 (cf. [73, Corollary 2.9]).

The intersection pattern of components of $\mathcal{S}_{3,1} \otimes k$ seems fairly complicated. For example, let $\rho_2 \in \mathcal{V}_1(\mathbf{F}_{p^2})$, and let $T' := \pi^{-1}(\rho_2) \subset \mathcal{P}_{3,\eta}$ be the fiber above ρ_2. Then
$$\#\{x \in T' | a(\Psi(x)) = 3\} = p^2 + 1, \qquad (9.4.17)$$
and W_η is non-singular at every superspecial point $x \neq \Psi(T) \in T'$. However, such an x equals $(E^3 \otimes k, \mu)$ for some principal polarization μ and is therefore a singular point in the component $W_{\eta'}$ with $\eta' = p\mu$.

The number of irreducible components of $\mathcal{S}_{3,1} \otimes k$ was shown in [36, Theorem 6.7] to equal $H_3(p,1)$. This number was explicitly computed by Hashimoto in [24, Theorem 4]. Note that $H_3(2,1) = 1$, furthermore $H_3(p,1) > 1$ for $p > 2$, and $H_3(p,1) \approx p^6/(2^9 \cdot 3^4 \cdot 5 \cdot 7)$ for p large.

For the action of $\mathrm{Aut}(E^3 \otimes k, \eta)$ on $\mathcal{P}_{3,\eta}$, see Proposition 9.12 below.

Conclusion. Let Λ be a set of representatives of equivalence classes of polarizations η of $E^3 \otimes \bar{\mathbf{F}}_p$ satisfying $\ker(\eta) = E^3[p] \otimes \bar{\mathbf{F}}_p$. Then there is a one to one correspondence ψ between Λ and the set of irreducible components of $\mathcal{S}_{3,1} \otimes \bar{\mathbf{F}}_p$. Again denote by W_η the irreducible component corresponding to η under ψ. Then W_η is birationally equivalent to $\mathcal{P}_{3,\eta}/G_\eta$, where $\mathcal{P}_{3,\eta}$ is a $\mathbf{P}^1$-bundle over a Fermat curve and $G_\eta = \mathrm{Aut}(E^3 \otimes \bar{\mathbf{F}}_p, \eta)/\{\pm 1\}$. We have $\#(\Lambda) = H_3(p,1)$ and
$$\mathcal{S}_{3,1} \otimes \bar{\mathbf{F}}_p = \bigcup_{\eta \in \Lambda} W_\eta. \qquad (9.4.18)$$

Note that W_η has a singular point corresponding to $(E^3 \otimes \bar{\mathbf{F}}_p, \eta/p)$ (see (9.4.16)), and the tangent space at this point to W_η has dimension 6 (see [73, Proposition 2.3]).

9.5. Some other methods for the calculation.

When $g > 3$, there are many global equations for $\mathcal{P}_{g,\eta}$ (i.e. more than the difference of the number of variables and the dimension), and one can hardly see the structure of $\mathcal{P}_{g,\eta}$ from these equations. So we will write down local equations in the sequel.

For convenience we will also use the language of Dieudonné modules (see 11.3 for an explanation).

9.6. Example, $g = 4$.

When $g = 4$, we first see that $\mathcal{V}_2$ is isomorphic to the non-singular surface $S \subset \mathbf{P}^3$ defined by (see 9.3.2))

$$a^{p^2}b - ab^{p^2} + c^{p^2}d - cd^{p^2} = 0. \tag{9.6.1}$$

Let x, y, z, u be the corresponding generators of the skeleton of $M_3 = A_{1,1}^{\oplus 4}$ (satisfying $\langle x, F^4 y \rangle = \langle z, F^4 u \rangle = 1$).

We consider an open neighborhood of a point $(a, b, c, d) \in S$, where a, b, c, d are linearly independent over $\mathbf{F}_{p^2}$. The corresponding Dieudonné module M_2 at (a, b, c, d) is generated by Fx, Fy, Fz, Fu and $v = \tilde{a}x + \tilde{b}y + \tilde{c}z + \tilde{d}u$, where $\tilde{a}, \tilde{b}, \tilde{c}, \tilde{d}$ are liftings of a, b, c, d in $W = W(k)$ respectively. To give an M_1 is equivalent to giving a vector $w = \tilde{r}v + \tilde{s}Fx + \tilde{t}Fy$ ($\tilde{r}, \tilde{s}, \tilde{t} \in W$, not all in pW) such that

$$\langle w, Fw \rangle \in W \tag{9.6.2}$$

or explicitly

$$rt^p a - rs^p b + sr^p b^p - tr^p a^p = 0 \tag{9.6.3}$$

where r, s, t are the images of $\tilde{r}, \tilde{s}, \tilde{t}$ in $W/pW \cong k$ respectively. Therefore we get two irreducible components $\mathcal{V}_{11}$ and $\mathcal{V}_{12}$ of $\mathcal{V}_1$, where $\mathcal{V}_{11}$ is defined by

$$t^p a - s^p b + sr^{p-1} b^p - tr^{p-1} a^p = 0, \tag{9.6.4}$$

hence $\mathcal{V}_{11} \to \mathcal{V}_2$ has fiber dimension 1, and $\mathcal{V}_{12}$ is defined by $r = 0$, hence it is a $\mathbf{P}^1$-bundle over $\mathcal{V}_2$. Therefore $\dim(\mathcal{V}_{11}) = \dim(\mathcal{V}_{12}) = 3$. (One can compare this with Remark 6.4. Here $r = 0$ means $\ker(F_{Y_3/S}) \subset \ker(Y_3 \to Y_1)$, in this case condition iii) in 6.2 automatically holds for $i = 1$.)

Since $\mathcal{V}_0 = \mathcal{P}_{4,\eta}$ is a $\mathbf{P}^1$-bundle over $\mathcal{V}_1$, we see that $\mathcal{V}_0$ also has two irreducible components $\mathcal{V}_{01}$ and $\mathcal{V}_{02}$ (both of dimension 4), where $\mathcal{V}_{02}$ is a $\mathbf{P}^1$-bundle over $\mathcal{V}_{12}$ and does not meet $\mathcal{P}'_{4,\eta}$. It is easy to check that the fiber of Y_0 over the generic point of $\mathcal{V}_{02}$ has a-number 2.

In general, if the fiber of Y_0 over the generic point of an irreducible component $\mathcal{V} \subset \mathcal{P}_{g,\eta}$ is not supergeneral, then we call $\mathcal{V}$ a "garbage component" of $\mathcal{P}_{g,\eta}$. Note that $\mathcal{V}$ is a garbage component iff it does not map surjectively to a component of $\mathcal{S}_{g,1}$. Note also that the generic point of a garbage component is not in $\mathcal{P}'_{g,\eta}$.

Thus $\mathcal{V}_{02}$ is a garbage component of $\mathcal{P}_{4,\eta}$.

On the other hand, when $p > 2$, we see that $\mathcal{V}_{11}$ is singular at a point with $r = 0$. Hence $\mathcal{V}_{01}$ is also singular.

By more calculation one can see that $\mathcal{P}_{4,\eta}$ is reduced.

9.7. A proof of Proposition 4.3.i) for $g = 4$.

We now show that $\mathcal{P}'_4$ is smooth over $\mathbf{F}_{p^2}$. This is simply an illustration of 11.3 for $g = 4$. By 3.9, it is enough to show that $\mathcal{P}'_{4,\eta}$ is non-singular for a special choice of η over $k = \bar{\mathbf{F}}_p$. We choose η such that for some decomposition $E^4 \otimes k \cong E_1 \times E_2 \times E_3 \times E_4$, we have $\eta = p(\eta'' \times \eta')$, where η' (resp. η'') is a polarization of $E_2 \times E_3$ (resp. $E_1 \times E_4$) such that $\ker(\eta') = (E_2 \times E_3)[F]$ (resp. $\ker(\eta'') = (E_1 \times E_4)[F]$).

Let $\{X_3 \to \dots \to X_0\}$ be the universal PFTQ over $\mathcal{P}'_{4,\eta}$. Let $U_i \subset \mathcal{P}'_{4,\eta}$ be the largest open subscheme such that $E_i \times U_i \to X_0 \times_{\mathcal{P}'_{4,\eta}} U_i$ is a closed immersion $(1 \leq i \leq 4)$. Then $\mathcal{P}'_{4,\eta} = \bigcup_i U_i$. By symmetry it is enough to show U_1 is non-singular. For convenience we denote $X_0 \times_{\mathcal{P}'_{4,\eta}} U_1$ simply by X_0.

Since $E_1 \times U_1 \to X_0$ is a closed immersion, its dual

$$X_0 \cong X_0^t \to E_1^t \times U_1 \cong (E_4/E_4[F^3]) \times U_1 \tag{9.7.1}$$

is smooth. Therefore the projections $X_i \to (E_4/E_4[F^{3-i}]) \times U_1$ $(0 \leq i \leq 3)$ are all smooth. Let $X_i'' = H_1(C_.^{i+1})$ $(i = 0,1)$, where $C_.^i$ is the complex

$$C_.^i : \quad E_1 \times U_1 \to X_i \to (E_4/E_4[F^{3-i}]) \times U_1. \tag{9.7.2}$$

Then one sees that $\{X_1'' \to X_0''\}$ is a PFTQ with respect to η'. This induces a morphism

$$\psi : U_1 \to \mathcal{P}_{2,\eta'} \cong \mathbf{P}^1. \tag{9.7.3}$$

It is enough to show ψ is smooth.

We first decompose ψ. Let $\{X_1' \to X_0'\}$ be the universal PFTQ over $\mathcal{P}_{2,\eta'}$ and $G' = \ker(X_1' \to X_0')$. Let $\mathfrak{U}_m$ $(0 \leq m \leq 3)$ be the category of sequences of isogenies $\{Y_3 \to \dots \to Y_m\}$ of polarized abelian schemes (Y_i, η_i) over some $\mathcal{P}_{2,\eta'}$-scheme S such that

i) $Y_3 = E^4 \times S$, with $\eta_3 = \eta \times \mathrm{id}_S$;
ii) $\ker(Y_i \to Y_{i-1})$ is a flat α-group of α-rank i $(m \leq i \leq 3)$;
iii) $\ker(Y_3 \to Y_i) = \ker(Y_3 \to Y_m) \cap Y_3[F^{3-i}]$ $(m < i \leq 3)$;
iv) $\ker(\eta_i) \subset X_i[F^i]$ $(m \leq i \leq 3)$;
v) $E_1 \times S \to Y_m$ is a closed immersion, and there are induced isomorphisms $\phi_i : H_1(C_.^{i+1}) \cong X_i' \times_{\mathcal{P}_{2,\eta'}} S$ $(m - 1 \leq i \leq 1)$, where $C_.^i$ is the complex

$$C_.^i : \quad E_1 \times S \to X_i \to (E_4/E_4[F^{3-i}]) \times S; \tag{9.7.4}$$

vi) (for $m < 3$ only) letting $G \subset Y_2[F]$ be the inverse image of $G' \times_{\mathcal{P}_{2,\eta'}} S$ in $Y_2[F]$ under

$$G' \times_{\mathcal{P}_{2,\eta'}} S \subset X_1'[F] \times_{\mathcal{P}_{2,\eta'}} S \hookrightarrow Y_2[F]/E_1[F] \times S \tag{9.7.5}$$

induced by ϕ_1 in v), we have $G^{(p)} \subset Y_2^{(p)}[F] \cap \ker(V : Y_2^{(p)} \to Y_2)$.

Let $\mathcal{U}_m$ be the fine moduli scheme of $\mathfrak{U}_m$. Then clearly $\mathcal{U}_0 \cong U_1$ and $\mathcal{U}_3 \cong \mathcal{P}_{2,\eta'}$. Furthermore, the truncations induce morphisms $\psi_i : \mathcal{U}_i \to \mathcal{U}_{i+1}$ $(0 \leq i \leq 2)$, and $\psi = \psi_2 \circ \psi_1 \circ \psi_0$. Hence it is enough to show each ψ_i is smooth. By 9.3.ii), we see ψ_0

is a line bundle (it is not a $\mathbf{P}^1$-bundle because of the open condition v)). It remains to check the smoothness of ψ_1 and ψ_2.

First we consider ψ_2. For a given $\{S; Y_3\} \in \mathrm{Ob}(\mathfrak{U}_3)$, let $G_1 = \ker(Y_3 \to Y_3^t) = Y_3[F^3]$ and $G_2 = Y_3[F]$. Note that G_2 is an α-group, and we denote by $\mathcal{F}$ the α-sheaf of G_2.

To extend $\{S; Y_3\}$ to an object of $\mathfrak{U}_2$, we need to find an α-subgroup $G_3 \subset G_2$ of α-rank 3, or equivalently a nowhere zero section s of $\mathcal{F}$. Condition v) simply says the s_1-coordinate of s is non-zero. Hence we can assume

$$s = s_1 + x_1 s_2 + x_2 s_3 + x s_4. \tag{9.7.6}$$

Let $G_4 = E_1[F] \times S$. Then $G_4 \subset Y_2 = Y_3/G_3$, and the projection $Y_2 \to Y_3/Y_3[F]$ gives an exact sequence

$$0 \to G_4 \to Y_2[F] \to G_5 \to 0, \tag{9.7.7}$$

where $G_5 = (E_2 \times E_3 \times E_4)[F]^{(p)} \times S$.

We check condition vi). Let $G_6 = Y_2^{(p)}[F] \cap \ker(V : Y_2^{(p)} \to Y_2)$. Then $G_7 = G_6/G_4^{(p)}$ is a subgroup scheme of $G_5^{(p)}$ by (9.7.7). It is easy to see that the ideal sheaf of $G_7 \hookrightarrow G_5^{(p)}$ is generated by the section $F^* s^{(p)} - V^* s = s^{(p^2)} - s$ of the α-sheaf of $G_5^{(p)}$. On the other hand G' is defined by the section $y_1 s_2^{(p)} + y_2 s_3^{(p)}$ of the α-sheaf $\mathcal{F}'$ of $X_1'[F]$, where y_1, y_2 are the homogeneous coordinates of $\mathcal{P}_{2,n'} \cong \mathbf{P}^1$. Hence vi) is equivalent to that the restriction of $s^{(p^2)} - s$ to $\mathcal{F}' \otimes_{\mathcal{O}_{\mathfrak{U}_3}} \mathcal{O}_S$ is proportional to $y_1 s_2^{(p)} + y_2 s_3^{(p)}$, or explicitly

$$(x_1^{p^2} - x_1) y_2 = (x_2^{p^2} - x_2) y_1. \tag{9.7.8}$$

Next we check condition iv). Since G_2^D is a quotient group scheme of $G_1^D \cong G_1$ and $\ker(G_1 \to G_2^D) = G_1[F^2]$, we have an induced isomorphism $f : G_2^D \to G_2^{(p^2)}$, which is equivalent to an $\mathcal{O}_S$-linear map $\mathcal{F}^{(p^2)} \to \mathcal{F}^\vee$, or equivalently an $\mathcal{O}_S$-bilinear form $\langle,\rangle : \mathcal{F} \otimes_{\mathcal{O}_S} \mathcal{F}^{(p^2)} \to \mathcal{O}_S$. Take a generator s_i of the α-sheaf of $E_i[F]$ for each i. Then s_1, s_2, s_3, s_4 can be viewed as a set of generators of $\mathcal{F}$. We can choose s_1, s_2, s_3, s_4 such that

$$\langle s_1, s_4^{(p^2)} \rangle = -\langle s_4, s_1^{(p^2)} \rangle = \langle s_2, s_3^{(p^2)} \rangle = \langle s_3, s_2^{(p^2)} \rangle = 1, \tag{9.7.9}$$

and we have

$$\langle s_1, s_2^{(p^2)} \rangle = \langle s_1, s_3^{(p^2)} \rangle = \langle s_4, s_2^{(p^2)} \rangle = \langle s_4, s_2^{(p^2)} \rangle = 0,$$
$$\langle s_i, s_i^{(p^2)} \rangle = 0 \quad (1 \leq i \leq 4). \tag{9.7.10}$$

Let $G_8 = G_2/G_3$. Then G_8^D is a subgroup scheme of G_2^D. Let $\phi : G_8^D \to G_8^{(p^2)}$ be the composition of the inclusion $G_8^D \hookrightarrow G_2^D$, f and the projection $G_2^{(p^2)} \twoheadrightarrow G_8^{(p^2)}$. Then iv) is equivalent to $\phi = 0$, and this is then equivalent to $\langle s, s^{(p^2)} \rangle = 0$, or explicitly

$$x^{p^2} - x + x_1 x_2^{p^2} - x_2 x_1^{p^2} = 0. \tag{9.7.11}$$

62

Note that we also have $G_2^D \cong G_1/G_1[p]$, which induces another bilinear form $\langle,\rangle_1 : \mathcal{F} \otimes_{O_S} \mathcal{F} \to O_S$. We automatically have $\langle s,s \rangle_1 = 0$ since $\langle,\rangle_1$ is alternating. Therefore we have $\ker(Y_2 \to Y_2^t) \subset Y_2[p]$ for any choice of G_3.

We see that $\mathfrak{U}_2 \to \mathfrak{U}_3$ is defined by variables x_2, x_3, x with defining relations (9.7.8) and (9.7.11), hence ψ_2 is smooth.

Finally we consider ψ_1. Assume we are given an object $\{S; Y_3 \to Y_2\}$ of $\mathfrak{U}_2$. Let $G_9 = Y_2[F]$ and $G_{10} = \ker(Y_2 \to Y_2^t)$. Then condition vi) says that we have an α-group $G \subset G_9$ of α-rank 2. On the other hand, condition iv) (for $i = 2$) says $G_{10} \subset Y_2[F^2]$, and the above note says $G_{10} \subset Y_2[p]$, hence $\mathrm{coker}(G_9 \to G_{10})$ has Verschiebung 0. Therefore

$$\ker(G_{10} \cong G_{10}^D \to G_9^D) \subset G_{10}[F] = G_9. \tag{9.7.12}$$

Thus we have an induced homomorphism $\phi : G_9^D \to G_9^{(p)}$. It is easy to see that ϕ^D induces a homomorphism $\Phi. : D'. \to D.$ of the following two complexes

$$D'. : \quad E_1[F] \times S \hookrightarrow (G_9^D)^{(p)} \twoheadrightarrow E_4[F]^{(p)} \times S \tag{9.7.13}$$

and

$$D. : \quad E_1[F] \times S \hookrightarrow G_9 \twoheadrightarrow E_4[F]^{(p)} \times S. \tag{9.7.14}$$

Note that Φ_0 and Φ_2 are isomorphisms and $H_1(\Phi.) = 0$. Hence ϕ^D has a flat image $G_{11} \subset G_9$, which is an α-group of α-rank 2.

Note that $G \cap G_{11} = E_1[F] \times S$, hence G and G_{11} together generate an α-group $G_{12} \subset G_9$ of α-rank 3. Let $\mathcal{F}'$ be the α-sheaf of G_{12}. Locally we can lift s_1 to a section s_1' of $\mathcal{F}'$. Locally we also take a section s' of $\mathcal{F}'$ which lifts a generator of the α-sheaf of $G' \times_{U_3} S$. Thus $\mathcal{F}'$ is locally generated by $s_1', s', s_4^{(p)}$.

To extend $\{S; Y_3 \to Y_2\}$ to an object of $\mathfrak{U}_1$, we need to find a subgroup scheme $G_{13} \subset G_9$ which is an α-group of α-rank 2 (and $Y_1 = Y_2/G_{13}$). We first show it is necessary that $G_{13} \subset G_{12}$. Indeed, since $G_{13} \cap E_1[F] \times S = 0$ by condition v), it is enough to show that the image G_{14} of G_{13} in $G_9/E_1[F] \times S$ is equal to $G_{12}/E_1[F] \times S$. Condition v) requires that $G' \times_{U_3} S \cong G/E[F] \times S \subset G_{14}$. On the other hand $E_1[F] \times S \subset Y_1$ and the above note gives a subgroup scheme

$$E_1^t[F]^{(p)} \times S \cong E_4[F]^{(p)} \times S \hookrightarrow G_9/E_1[F] \times S \tag{9.7.15}$$

which maps to 0 in $Y_1/E_1 \times S$ by the dual of iv). Hence we have $E_4[F]^{(p)} \times S \cong G_{11}/E_1[F] \times S \subset G_{14}$.

It reduces to finding a section $s = s_1' + x_1 s' + x s_4^{(p)}$ of $\mathcal{F}'$. It remains to check condition iv). As in the case of ψ_2, we have an induced homomorphism $G_{12}^D \to G_{12}^{(p)}$ which is equivalent to an O_S-bilinear form $\langle,\rangle_2 : \mathcal{F}' \otimes_{O_S} \otimes \mathcal{F}'^{(p)} \to O_S$, and iv) is equivalent to

$$\langle s, s^{(p)} \rangle_2 = 0. \tag{9.7.16}$$

Note that we have

$$\begin{aligned}
&\langle s_1', s_4^{(p^2)} \rangle_2 = -\langle s_4^{(p)}, s_1'^{(p)} \rangle_2 = 1, \\
&\langle s', s_4^{(p^2)} \rangle_2 = \langle s_4^{(p)}, s'^{(p)} \rangle_2 = \langle s_4^{(p)}, s_4^{(p^2)} \rangle_2 = \langle s', s'^{(p)} \rangle_2 = 0
\end{aligned} \tag{9.7.17}$$

But $c = \langle s_1', s'^{(p)}\rangle_2$, $c' = \langle s', s_1'^{(p)}\rangle_2$ and $d = \langle s_1', s_1'^{(p)}\rangle_2$ may not equal 0 in general. Thus we can write (9.7.16) explicitly

$$x^p - x + cx_1 + c'x_1^p + d = 0. \tag{9.7.18}$$

Therefore $\mathcal{U}_1 \to \mathcal{U}_2$ is locally given by variables x_1, x with the defining relation of the form (9.7.18). Hence ψ_1 is smooth.

Remark. Let $\mathcal{V}_m' \subset \mathcal{V}_m$ (see Example 9.6) be the open subscheme representing sequences $\{Y_3 \to ... \to Y_m\}$ (over a k-scheme S) satisfying i-iv) above and $E_1 \times S \hookrightarrow Y_m$. Then we have $\mathcal{V}_0' \cong \mathcal{U}_0$ and $\mathcal{V}_1' \cong \mathcal{U}_1$. On the other hand, we have an induced morphism $\mathcal{U}_2 \to \mathcal{V}_2 \times \mathcal{P}_{2,\eta'}$ which is not an isomorphism because $\mathcal{V}_2 \times \mathcal{P}_{2,\eta'}$ represents isogenies $\{Y_3 \to Y_2\}$ satisfying i-v) but not vi). Condition vi) guarantees that an extension $\{Y_3 \to Y_2 \to Y_1\}$ of $\{Y_3 \to Y_2\}$ satisfies v) also.

9.8. Garbage components for large g.

Example. When $g = 5$, by the same way of calculation we see that $\mathcal{P}_{g,\eta}$ has a garbage component of dimension 6, which is equal to $\dim(\mathcal{S}_{5,1})$. When $g > 5$, we even have a garbage component of dimension $> [g^2/4]$.

9.9. The subsets defined by a-numbers.

For any $n > 0$, the points of $\mathcal{S}_{g,d}$ representing abelian varieties with a-number $\geq n$ form a Zariski closed subset, which will be denoted by $\mathcal{S}_{g,d}(a \geq n)$. For example, $\mathcal{S}_{g,d}(a \geq g)$ is the set of superspecial points, and $\mathcal{S}_{g,1}(a \geq 2)$ is a divisor, as will be shown in Corollary 10.3.

We now study $\mathcal{S}_{4,1}(a \geq n)$. There are two kinds of irreducible components in $\mathcal{S}_{4,1}(a \geq 2)$:

a) Let μ be a polarization of E^4 such that $\ker(\mu) = E^4[p]$. Consider sequences of isogenies of polarized abelian varieties

$$E^4 \to Y_1 \to Y_0 \quad (\ker(E^4 \to Y_1) \cong \alpha_p^3, \ \ker(Y_1 \to Y_0) \cong \alpha_p) \tag{9.9.1}$$

where the polarization of E^4 is μ. Such sequences admit a fine moduli scheme U_μ which is isomorphic to a $\mathbf{P}^1$-bundle of the hypersurface

$$X_1^{p+1} + X_2^{p+1} + X_3^{p+1} + X_4^{p+1} = 0 \tag{9.9.2}$$

in $\mathbf{P}^3$. The image of U_μ in $\mathcal{S}_{4,1}(a \geq 2)$ is an irreducible component of dimension 3, and there are $H_4(p, 1)$ irreducible components of this kind.

b) Let μ be a polarization of E^4 such that $\ker(\mu) = E^4[F]$. Consider isogenies of polarized abelian varieties

$$E^4 \to Y_0 \quad (\ker(E^4 \to Y_1) \cong \alpha_p^2) \tag{9.9.3}$$

where the polarization of E^4 is μ. Such isogenies admit a fine moduli scheme $\mathcal{T}_\mu$ which is isomorphic to the subscheme of the Grassmannian $\mathrm{Grass}_{4,2}$ consisting of points representing isotropic subspaces of $\{k^{\oplus 4}, \langle , \rangle\}$, where $\langle , \rangle$ is a non-degenerate alternating form. The image of T_μ in $\mathcal{S}_{4,1}(a \geq 2)$ is an irreducible component of dimension 3, and there are $H_4(1,p)$ irreducible components of this kind.

9.10. Supersingular Dieudonné modules with a-number $g - 1$.

Next we study $\mathcal{S}_{g,1}(a \geq g - 1)$ (in particular $\mathcal{S}_{4,1}(a \geq 3)$). We make use of the following result.

Lemma. *Let M be a principally quasi-polarized supersingular Dieudonné module of genus g over $W(k)$ with $a(M) = g - 1$. Then there is a decomposition $M = N \oplus N'$, where N' is a principally quasi-polarized superspecial Dieudonné module, and N is a principally quasi-polarized Dieudonné module of genus $2r$ $(r \leq g/2)$ such that $S_0 N = F S^0 N$.*

Proof. By $a(M) = g-1$ we have $FS^0 M \subset M$ (see [45, p.337]). Hence by Proposition 6.1 we have a decomposition $S^0 M = N_0 \oplus N'$, where N' is a principally quasi-polarized superspecial Dieudonné module, and N_0 is a quasi-polarized superspecial Dieudonné module such that $N_0^t = F N_0$. Let $N = M \cap N_0$ and $r = \dim_k(N/N_0^t)$. Then $M = N \oplus N'$ and $N^t = N$. Finally, since $r = \dim_k(N_0/N^t) = \dim_k(N_0/N_0^t) - r$, we see that $g(N) = g(N_0) = 2r$. Q.E.D.

9.11. The structure of $\mathcal{S}_{g,1}(a \geq g - 1)$.

Proposition. *Let $k = \bar{\mathbb{F}}_p$. For any $0 < r \leq [g/2]$ and any polarization μ of $E^g \otimes k$ such that $\ker(\mu) \cong \alpha_p^{2r}$, denote by $\mathcal{T}_\mu$ the fine moduli scheme of isogenies $\rho : E^g \otimes k \to Y$ of polarized abelian varieties satisfying*

$$\ker(\rho) \cong \alpha_p^r \subset \ker(\mu), \qquad (9.11.1)$$

where the polarization of $E^g \otimes k$ is μ (hence Y is principally polarized). Denote by $T_\mu \subset \mathcal{T}_\mu$ the locally closed subset of points whose corresponding Y has $a(Y) = g-1$ (with reduced induced scheme structure).

i) The induced morphism $T_\mu \to \mathcal{S}_{g,1} \otimes k$ is generically finite to one, and T_μ is irreducible of dimension r.

ii) The induced morphism

$$\Psi_0 : \coprod_{\ker(\mu) \cong \alpha_p^{2[g/2]}} T_\mu \to \mathcal{S}_{g,1}(a \geq g - 1) \otimes k \qquad (9.11.2)$$

is surjective and gives a one to one correspondence between the set of irreducible components of $\mathcal{S}_{g,1}(a \geq g - 1) \otimes k$ and the set of equivalence classes of μ such that $\ker(\mu) \cong \alpha_p^{2[g/2]}$.

iii) Every irreducible component of $\mathcal{S}_{g,1}(a \geq g-1)$ has dimension $[g/2]$, and the number of irreducible components of $\mathcal{S}_{g,1}(a \geq g - 1) \otimes k$ is equal to $H_g(1,p)$.

Proof. i) To give an isogeny $\rho : E^g \otimes k \to Y$ satisfying (9.11.1) is equivalent to giving a totally isotropic subspace of dimension r of the α-sheaf $\mathcal{F}$ of $\ker(\mu)$, or an

$r \times r$ symmetric matrix $C = (c_{ij})$ over k under a choice of standard basis of $\mathcal{F}$. By an easy calculation of Dieudonné modules, one sees that $a(Y) = g - 1$ iff the corresponding C satisfies

$(*)$ $C - C^{(p^2)} = (c_{ij} - c_{ij}^{p^2})$ has rank 1.

By the symmetricity of C, $(*)$ is equivalent to $r(r-1)/2$ local equations on c_{ij} $(1 \le i, j \le r)$. Hence every irreducible component of T_μ has dimension $\ge r(r+1)/2 - r(r-1)/2 = r$.

Let $T'_{2r} \subset T_{2r}$ (in Proposition 9.3) be the set of points whose coordinates are linearly independent over F_{p^2}. Then T'_{2r} is open dense in T_{2r} by Proposition 9.3, hence has dimension r. For any $(a_1, ..., a_{2r}) \in T'_{2r}$, under a choice of standard basis of $\mathcal{F}$, the subspace of $\mathcal{F}$ generated by

$$(a_1^{p^{2n}}, ..., a_{2r}^{p^{2n}}) \quad (0 \le n < r) \tag{9.11.3}$$

is totally isotropic of dimension r by (9.3.2), hence gives a minimal isogeny ρ as in (9.11.1). This gives a morphism $\phi_\mu : T'_{2r} \to T_\mu$ which is easily seen to be set-theoretically injective. Conversely, if the isogeny ρ in (9.11.1) is minimal, then ρ is represented by a point in $\text{im}(\phi_\mu)$. Combining this with the fact that every irreducible component of T_μ has dimension $\ge r$ (as shown above), we see that ϕ_μ is generically surjective and T_μ is irreducible of dimension r.

Furthermore, if ρ is minimal, then μ is uniquely determined by the polarization of Y. Hence $T_\mu \to \mathcal{S}_{g,1} \otimes k$ is generically finite to one.

ii) By Lemma 9.10, the morphism Ψ_0 in (9.11.2) is surjective. We have also seen that $\Psi_0(T_\mu)$ determines the equivalence class of μ, hence Ψ_0 gives a one to one correspondence between the irreducible components of $\mathcal{S}_{g,1}(a \ge g-1) \otimes k$ and the equivalence classes of μ.

iii) By i) and ii) we see that every irreducible component of $\mathcal{S}_{g,1}(a \ge g-1) \otimes k$ has dimension $[g/2]$, and the number of irreducible components of $\mathcal{S}_{g,1}(a \ge g-1) \otimes k$ is equal to the number of equivalence classes of μ such that $\ker(\mu) \cong \alpha_p^{2[g/2]}$, which is equal to $H_g(1, p)$ by Corollary 4.8.iii).   Q.E.D.

9.12. The action of the automorphism group of a polarization η on $\mathcal{P}_{g,\eta}$.

We study the action of $\text{Aut}(E^g \otimes k, \eta)$ on $\mathcal{P}_{g,\eta}$ for any $g > 1$.

Proposition. *Let $g > 1$ and η be a polarization of $E^g \otimes k$ such that $\ker(\eta) = E^g[F^{g-1}] \otimes k$.*

i) *If g is odd and $\eta = p^{(g-1)/2}\mu^g$ for some principal polarization μ of (a choice of) E, then the group $\text{Aut}(E^g \otimes k, \eta)$ is isomorphic to the subgroup of $GL_g(\mathcal{O})$ consisting of matrices T such that each row of T has one entry in $\mathcal{O}^\times$ with the other entries$= 0$, hence we have*

$$\text{Aut}(E^g \otimes k, \eta) \cong (\mathcal{O}^\times)^g \rtimes S_g. \tag{9.12.1}$$

ii) *If $g \ne 3$ or $p > 2$, then the action of $\text{Aut}(E^g \otimes k, \eta)/\{\pm 1\}$ on $\mathcal{P}'_{g,\eta}$ is generically free.*

66

iii) *When $g = 3$ and $p = 2$, (9.12.1) holds since there is only one equivalence class of η. The stabilizer of the generic point of $\mathcal{P}_{3,\eta}$ under the action of $\operatorname{Aut}(E^3 \otimes k, \eta)$ is isomorphic to $\{\pm 1\}^3 \subset (\mathcal{O}^\times)^3$ under (9.12.1), and the degree of $\mathcal{P}_{3,\eta} \to \mathcal{S}_{3,1}$ is $2^7 3^4$.*

Proof. From [31, Proposition 2.8] we see that for any choice of a principal polarization μ_0 of $E^g \otimes k$, there is an isomorphism

$$\operatorname{Aut}(E^g \otimes k, \eta) \cong \{T \in GL_g(\mathcal{O}) | \bar{T}^t A T = A\} \tag{9.12.2}$$

via $\operatorname{Aut}(E^g \otimes k, \eta) \subset \operatorname{Aut}(E^g \otimes k) \cong GL_g(\mathcal{O})$, where $A = \mu_0^{-1} \circ \eta$ and $\bar{T}$ is the Rosati involution of T with respect to μ_0. When g is odd, we can take $\mu_0 = \eta / p^{(g-1)/2}$, hence

$$\operatorname{Aut}(E^g \otimes k, \eta) \cong \{T \in GL_g(\mathcal{O}) | \bar{T}^t T = I_g\}. \tag{9.12.3}$$

Write $T = (a_{ij}) \in GL_g(\mathcal{O})$. When $\mu_0 = \mu^g$, we have $\bar{T}^t = (\bar{a}_{ji})$, where $\bar{a}_{ji}$ is the conjugate of $a_{ji} \in \mathcal{O}$ (i.e. the Rosati involution with respect to μ). Note that $\bar{a}_{ij} a_{ij}$ is a positive integer unless $a_{ij} = 0$. Hence $\bar{T}^t T = I_g$ is equivalent to that each row of T has one entry in $\mathcal{O}^\times$ with the other entries$=0$. Therefore we have an exact sequence

$$(\mathcal{O}^\times)^g \hookrightarrow \operatorname{Aut}(E^g \otimes k, \eta) \twoheadrightarrow S_g. \tag{9.12.4}$$

This proves i).

Next we prove ii). We have already seen the case $g = 2$ in 9.2, hence we assume $g > 2$ in the following. Note that $\operatorname{Aut}(E^g \otimes k, \eta)$ is a finite group.

Let $T_0 \subset \mathcal{P}'_{g,\eta}$ be the Zariski open subset of points representing PFTQs with supergeneral end. Let $x \in T_0$ represent a PFTQ $\{X_{g-1} \to \dots \to X_0\}$ with respect to η (with $a(X_0) = 1$). Then η induces a quasi-polarization $\langle , \rangle$ on $M_{g-1} = D(X_{g-1}) \cong A_{1,1}^{\oplus g}$ and we have $\langle M_0, M_0 \rangle \subset W$, where $M_0 = D(X_0)$. Since $a(M_0) = 1$, by Fact 5.6.ii) we have $M_0 = Av$ for some $v \in M_0$. Choose generators $x_1, \dots, x_g$ of the skeleton of M_{g-1} (see 5.7). Then we can write

$$v = (a_1 + b_1 F)x_1 + \dots + (a_g + b_g F)x_g \tag{9.12.5}$$

where $a_i, b_i \in W$ $(1 \le i \le g)$.

Let $\phi \in \operatorname{Aut}(E^g \otimes k, \eta)$. Then ϕ induces an automorphism $D(\phi)$ of M_{g-1} which preserves $\langle , \rangle$. Thus $D(\phi)$ can be expressed as an H-matrix $(\alpha_{ij} + \beta_{ij} F)$ $(\alpha_{ij}, \beta_{ij} \in W(\mathbb{F}_{p^2}))$ with respect to the generators $x_1, \dots, x_g$ (see (5.7.1) for the definition of H).

Suppose $\phi(x) = x$. Then $D(\phi)(M_0) = M_0$, i.e. $D(\phi)(v) \in Av$. Hence there exists $c \in k$ such that

$$\sum_i \bar{\alpha}_{ij} \bar{a}_i = c \bar{a}_j \quad (1 \le j \le g), \tag{9.12.6}$$

where $\bar{\alpha}_{ij}, \bar{a}_i$ are respectively the images of α_{ij}, a_i in k under the projection $W \to W/pW \cong k$. If v is general enough, we have $\bar{\alpha}_{ij} = c \delta_{ij}$. Since $D(\phi)$ preserves $\langle , \rangle$, we have $c^2 = 1$ (hence $c = \pm 1$) when g is even, and $c^{p+1} = 1$ when g is odd. Take a lifting $\tilde{c} \in W$ of c.

By $D(\phi)(v) \in Av$ we get

$$\sum_{i,j} (a_i + b_i F)(\tilde{c} \delta_{ij} + \beta_{ij} F)x_j \equiv \tilde{c}v + c_1 Fv + c_2 Vv \pmod{F^2 M_{g-1}} \tag{9.12.7}$$

67

for some $c_1, c_2 \in W$. By writing down the explicit equations one sees (9.12.7) is a non-trivial algebraic condition unless $c = \pm 1$ and $\beta_{ij} \in pW$ for all i, j. Therefore either

$$D(\varphi) \equiv \pm \text{id} \pmod{pH}, \tag{9.12.8}$$

or there is a non-empty Zariski open subset $U'_\phi \subset T_0$ such that $\phi(x) \neq x$ for any $x \in U'_\phi$.

Repeating the above argument inductively we can show that either

$$D(\phi) \equiv \pm \text{id} \pmod{F^{g-1}H}. \tag{9.12.9}$$

or there is a non-empty Zariski open subset $U_\phi \subset T_0$ such that $\phi(x) \neq x$ for any $x \in U_\phi$.

Now we use [46, Lemma 2.5 and Remark 2.6], which in particular gives:

(*) Let $\phi \in \text{Aut}(E^g \otimes k, \eta)$. If $p > 2$ and $E^g[p] \otimes k \subset \ker(\phi - \text{id})$, or if $p = 2$ and $E^g[F^3] \otimes k \subset \ker(\phi - \text{id})$, then $\phi = \text{id}$.

Note that (9.12.9) is equivalent to

$$E^g[F^{g-1}] \otimes k \subset \ker(\varphi \mp \text{id}), \tag{9.12.10}$$

hence implies $\phi = \pm \text{id}$ if $g > 3$ or $p > 2$.

When $g > 3$ or $p > 2$, let $U = \bigcap_{\phi \neq \pm \text{id}} U_\phi$. Then the stabilizer of any $x \in U$ in $\text{Aut}(E^g \otimes k, \eta)$ is $\{\pm \text{id}\}$. This proves ii).

Finally we consider the case $g = 3$, $p = 2$. In this case there is only one equivalence class of η because $H_3(2, 1) = 1$ (see [24, Theorem 4]). Hence we may assume $\eta = 2\mu^3$, where μ is a principal polarization of $E \otimes k$. Therefore $\text{Aut}(E^3 \otimes k, \eta)$ is isomorphic to the group of 3×3-matrices each row of which has an entry in $\mathcal{O}^\times$ with other entries $= 0$. Note that $\mathcal{O}^\times$ is a non-commutative group of order 24 (isomorphic to a semi-direct product of the quaternion group with $\mathbb{Z}/3\mathbb{Z}$). Since $\mathcal{P}_{3,\eta} \cong \mathcal{P}_3 \otimes k$ which only depends on $E^3[2] \otimes k$ (see 3.9), if ϕ acts trivially on $E^3[2] \otimes k$, then it acts trivially on $\mathcal{P}_{3,\eta}$. The converse also holds by the argument of ii).

If ϕ acts trivially on $E^3[2] \otimes k$, then we can write $\phi = \text{id} + 2\psi$, hence φ corresponds to a diagonal matrix $\text{diag}(\alpha_1, \alpha_2, \alpha_3) \in GL_g(\mathcal{O})$ and we can write $\alpha_i = 1 + 2\beta_i$ ($1 \leq i \leq 3$). Since the order of α_i divides 12, it is easy to check that $\alpha_i = \pm 1$. Hence ϕ acts trivially on $\mathcal{P}_{3,\eta}$ iff it corresponds to a diagonal matrix with ± 1 as its diagonal entries. This shows the first assertion of iii), and hence the degree of $\mathcal{P}_{3,\eta} \to \mathcal{S}_{3,1}$ is $2^7 3^4$.   Q.E.D.

9.13. Different automorphism groups of polarizations.

Remark. When g is even, the structure of $\text{Aut}(E^g \otimes k, \eta)$ (as a group) depends on η, as we have seen for $g = 2$ in 9.2 above. This is also the case when g is odd. For example, when $p = 17$, there are two supersingular elliptic curves E_1, E_2 over k with $j(E_1) = 8$ and $j(E_2) = 0$. We have $\text{Aut}(E_1) \cong \mathbb{Z}/2\mathbb{Z}$ and $\text{Aut}(E_2) \cong \mathbb{Z}/4\mathbb{Z}$. Take principal polarizations μ_1 of E_1 and μ_2 of E_2 respectively, and let $\eta_1 = p\mu_1^3$, $\eta_2 = p\mu_2^3$ which are polarizations of $E_1^3 \cong E_2^3$. Then by Proposition 9.12.i) we have $\text{Aut}(E_1^3, \eta_1) \not\cong \text{Aut}(E_2^3, \eta_2)$ (hence $\eta_1 \not\sim \eta_2$).

10. Main results on $\mathcal{S}_{g,d}$
(the non-principally polarized case)

In this chapter we collect the main propositions concerning the structure of $\mathcal{S}_{g,d}$ ($d > 1$) which will be proved in Chapter 11, and prove the main theorem on $\mathcal{S}_{g,d}$ using these propositions.

10.1. The moduli of rigid FTQs.

Proposition. *Let $\{Y_{g-1} \to \ldots \to Y_0\}$ be the universal FTQ over $\mathcal{Q}'_g$ (see Lemma 3.5).*

> i) *$\mathcal{Q}'_g$ is smooth (over F_p), geometrically integral and rational of dimension $g(g-1)/2$.*
> ii) *Every isomorphism class of supersingular abelian varieties of dimension g over k has a representative in $\mathcal{Q}'_g \otimes k$ (or in other words, every supersingular abelian variety is the end of a rigid FTQ).*
> iii) *There is a divisor $D \subset \mathcal{Q}'_g$ such that any geometric point s of $\mathcal{Q}'_g$ satisfies $a((Y_0)_s) > 1$ iff s is a geometric point of D.*

Without rigidity, ii) was already proved in [61, Theorem 2.2] (also cf. Proposition 4.1 and Remark 6.4).

10.2. A moduli of isogenies.

Remark. Because of Proposition 10.1.ii), we may view $\mathcal{Q}'_g$ as a "catalogue" of all supersingular abelian varieties of dimension g. But the disadvantage is that $\mathcal{Q}'_g$ is not proper when $g > 2$. We will see a way to compactify $\mathcal{Q}'_g$, namely:

Proposition. *Let $g > 1$, $n > 0$ and $\mathfrak{D}_{g,n}$ be the category of isogenies $E^g \times S \to Y$ of degree p^n, where S is an F_p-scheme.*

> i) *$\mathfrak{D}_{g,n}$ has a fine moduli scheme $\mathcal{D}_{g,n}$ which is projective over F_p.*
> ii) *The canonical morphism $\mathcal{Q}'_g \to \mathcal{D}_g := \mathcal{D}_{g,g(g-1)/2}$ induced by $Y_{g-1} \to Y_0$ (see Proposition 4.7) is an open immersion.*
> iii) *Every isomorphism class of a supersingular abelian variety has at least one but only a finite number of representatives in $\mathcal{D}_g \otimes k$.*

The proof will be given in 11.4.

By ii), we can identify $\mathcal{Q}'_g$ with an open subscheme of $\mathcal{D}_g$. We will denote by $\mathcal{D}'_g$ the closure of $\mathcal{Q}'_g$ in $\mathcal{D}_g$, with reduced induced structure.

For a general family of supersingular abelian varieties $Y \to S$ (of dimension g), there may not be an isogeny $E^g \times S \to Y$ at all (see examples in [45, p.345] and [69, Remark 10]). But we have the following theorem of Ogus (see [45, Lemma 4.1]).

Fact. *If S is normal, then there is an étale cover $S' \to S$ such that there is an isogeny $\rho : E^g \times S' \to Y \times_S S'$ (over S').*

By [45, Corollary 4.3], the least possible degree of ρ equals p^n for some n, and we can take ρ such that $\deg(\rho) = p^{g(g-1)/2}$.

In the following, when we say a supersingular abelian variety Y (of dimension g) is "general enough", we mean that there has been fixed a non-empty Zariski open subset $U \subset \mathcal{Q}'_g$ (viewed as an open subscheme of $\mathcal{D}_g$), and the representatives of Y in $\mathcal{D}_g$ are all in U. In this case we also say the Dieudonné module $D(Y)$ is general enough.

10.3. A result about $\mathcal{S}_{g,1}(a \geq 2)$.

Recall that in 9.9 we have defined $\mathcal{S}_{g,d}(a \geq n)$, i.e. the Zariski closed subset of $\mathcal{S}_{g,d}$ of points representing abelian varieties with a-number $\geq n$ (with reduced induced scheme structure). By Proposition 10.1.iii), Corollary 4.2 and Lemma 3.7 we get

Corollary. *For any $g > 1$, every irreducible component of $\mathcal{S}_{g,1}(a \geq 2)$ is of codimension 1 in $\mathcal{S}_{g,1}$.*

10.4. The representatives of a supersingular abelian variety in $\mathcal{A}_{g,d}$.

Next we study the following question.

Question. For which d, every isomorphism class of supersingular abelian varieties of dimension g has a representative in $\mathcal{A}_{g,d}$?

We will use the following

Proposition. *Let $\{Y, \eta_0\}$ be a polarized supersingular abelian variety over k and $\rho : E^g \otimes k \to Y$ be an isogeny. Let $\eta = \rho^t \circ \eta_0 \circ \rho$, the polarization of $E^g \otimes k$ induced by η_0.*

   i) If Y is general enough, then

$$\ker(\eta) \supseteq \ker(F_{E^g \otimes k}^{2g-3}) \tag{10.4.1}$$

   and

$$p^{2n} \,|\, \deg(\eta_0), \quad (n = [\frac{(g-1)^2}{2}]). \tag{10.4.2}$$

   ii) Conversely, for any isogeny $E^g \otimes k \to Y$ represented by a point in $\mathcal{D}'_g$ (see Proposition 10.2 above), any polarization η of $E^g \otimes k$ satisfying (10.4.1) induces a polarization η_0 of Y (satisfying (10.4.2) automatically). In particular, there is a polarization of Y of degree $p^{2[(g-1)^2/2]}$.

   iii) Let n be the integer such that $p^n \,||\, \deg(\eta_0)$. If $0 < n < g/2$, then $a(Y) > 1$.

The proof will be given in 11.12.

Remark. When g is odd, the equality of (10.4.1) does not hold, because the right hand side of (10.4.1) has rank $p^{g(2g-3)}$ which is not a square.

10.5. The main theorem on $\mathcal{S}_{g,d}$.

Theorem. Let $k = \bar{\mathbb{F}}_p$ and $g > 1$.

i) The following are equivalent:

a) $p^{[(g-1)^2/2]}|d$;

b) $\mathcal{S}_{g,d}$ has an irreducible component of dimension $g(g-1)/2$, and every isomorphism class of supersingular abelian varieties of dimension g has a representative in every irreducible component of $\mathcal{S}_{g,d} \otimes k$ of dimension $g(g-1)/2$;

c) Every isomorphism class of supersingular abelian varieties of dimension g has a representative in $\mathcal{A}_{g,d}$;

d) $\dim(\mathcal{S}_{g,d}) = g(g-1)/2$.

ii) If $d = p^{[(g-1)^2/2]}$, then the number of irreducible components of dimension $g(g-1)/2$ in $\mathcal{S}_{g,d} \otimes k$ is equal to $H_g(1,p)$.

Proof. i) Let V be an irreducible component of $\mathcal{S}_{g,d}$. Then we can find a family of polarized supersingular abelian varieties $\{Y, \eta_0\}$ over an integral base S, where $\deg(\eta_0) = d^2$, such that the induced morphism $S \to \mathcal{S}_{g,d}$ is finite onto V. By Fact 10.2, there is a finite epimorphism $T \to S$ such that there is an isogeny $\rho : E^g \times T \to Y \times_S T$. We can choose ρ so that it induces a rigid FTQ structure on the generic fiber of $Y \times_S T$ over T. Then ρ induces a morphism $h : T \to \mathcal{D}'_g$ (see Proposition 10.2 above), which is finite because every abelian variety has only a finite number of polarizations of degree d^2 up to equivalence. This shows that

$$\dim(V) \le \frac{g(g-1)}{2}, \qquad (10.5.1)$$

where the equality holds iff h is an epimorphism. By Proposition 10.1.ii), we see h is an epimorphism iff every isomorphism class of supersingular abelian varieties of dimension g has a representative in V. This shows b)$\Leftrightarrow$c)$\Leftrightarrow$d).

a)$\Rightarrow$c): Let η be a polarization of $E^g \otimes k$ of degree $d^2 p^{g(g-1)}$ satisfying (10.4.1). Then by Proposition 10.4.ii), we see η induces a morphism $\mathcal{D}'_g \otimes k \to \mathcal{S}_{g,d} \otimes k$. Hence every isomorphism class of supersingular abelian varieties of dimension g has a representative in $\mathcal{S}_{g,d}$.

c)$\Rightarrow$a): Take a supersingular abelian variety Y of dimension g which is general enough. Then c) says that Y has a polarization of degree d^2. Hence $p^{[(g-1)^2/2]}|d$ by (10.4.2).

ii) Let $d = p^{[(g-1)^2/2]}$. Let Y be a *supergeneral* abelian variety over k with a polarization η_0 of degree d^2 and an isogeny $\rho : E^g \otimes k \to Y$ of degree $g(g-1)/2$. Let $\eta = \rho^t \circ \eta_0 \circ \rho$. Suppose Y is general enough. Then (10.4.1) holds by Proposition 10.4.i). (In fact, $\ker(F^{2g-3}) = \ker(\eta)$ when g is even and $\ker(\eta)/\ker(F^{2g-3}) \cong \alpha_p$ when g is odd). Let $s \in \mathcal{S}_{g,d} \otimes k$ represent $\{Y, \eta_0\}$. Then s is in the image of the canonical morphism $\mathcal{D}'_g \otimes k \to \mathcal{S}_{g,d} \otimes k$ induced by η. By i), we see that

71

every irreducible component of $\mathcal{S}_{g,d} \otimes k$ of dimension $g(g-1)/2$ is the image of $\mathcal{D}'_g \otimes k \to \mathcal{S}_{g,d} \otimes k$ induced by some η as above. On the other hand, η is uniquely determined by η_0 up to equivalence. Therefore we get a one to one correspondence

$$\{\text{irreducible components of } \mathcal{S}_{g,d} \otimes k \text{ of dimension } g(g-1)/2\} \longleftrightarrow$$
$$\{\text{equivalence classes of polarizations } \eta \text{ of } E^g \otimes k \text{ of degree } d^2 p^{g(g-1)}$$
$$\text{satisfying } (10.4.1)\}$$

Let n_g be the number of equivalence classes of isogenies η of $E^g \otimes k$ of degree $d^2 p^{g(g-1)}$ satisfying (10.4.1). Then we see that the number of irreducible components of $\mathcal{S}_{g,d} \otimes k$ of dimension $g(g-1)/2$ is equal to n_g, and $n_g = H_g(1,p)$ by Corollary 4.4.ii). Q.E.D.

Remark. The case $g = 2$ of Theorem 10.5 is of course well-known, and 10.5.i) is also known for $g = 3$ (see [36, Remark 6.10]).

Note that $\mathcal{S}_g$ is contained in the very special locus $V_0 \subset \mathcal{A}_g \otimes \mathbf{F}_p$, hence by [48, Theorem 4.1], every irreducible component of $\mathcal{S}_g$ has dimension at most $g(g-1)/2$.

10.6. Components of different dimensions in $\mathcal{S}_{g,d}$.

Remark. When $d = p^{[(g-1)^2/2]}$ and $g > 3$, the locus $\mathcal{S}_{g,d}$ has an irreducible component of dimension $g(g-1)/2$ and has an irreducible component of dimension $< g(g-1)/2$. This can be shown as follows. Take a polarization η_0 of $Y = E^g$ which is a product of a principal polarization of E^{g-1} and a polarization of E of degree d^2. Let $x \in \mathcal{S}_{g,d}$ represent $\{Y, \eta_0\}$. Then x is not contained in an irreducible component of dimension $g(g-1)/2$. Indeed, if this were true, then by the proof of Theorem 10.5 we would have $\ker(\eta_0) \subset Y[F^{2g-2}]$, absurd when $g > 3$.

However, when $g = 2$, every irreducible component of $\mathcal{S}_{2,d}$ for any d has dimension 1 by [58, Theorem 4.1] (because when $g = 2$ we have: X is supersingular $\Leftrightarrow$ the p-rank of X is 0).

11. Proofs of the propositions on FTQs

In this chapter we give proofs of Propositions 10.1, 10.2 and 10.4. We also sketch a proof of Proposition 4.3, especially of the fact that $\mathcal{P}'_g$ is non-singular (recall that we only proved the Weak Form of Proposition 4.3 in Chapter 7). The proof of Proposition 10.1 uses a direct argument on group schemes, without using Dieudonné modules.

11.1. Proof of Proposition 10.1.

Proof of i). By [61, Theorem 2.4] it is enough to show the smoothness and the geometric integrality. Again we will use double induction.

Let $j_n : E \to E^g$ be the nth inclusion $(1 \leq n \leq g)$. Let $\mathfrak{Q}^n_g$ be the subcategory of $\mathfrak{Q}'_g$ consisting of objects $\{S; Y_{g-1} \to ... \to Y_0\}$ such that the homomorphism $E \times S \to Y_0$ induced by j_n is a monomorphism. Let $\mathcal{Q}^n_g$ be the fine moduli scheme of $\mathfrak{Q}^n_g$. Clearly $\mathcal{Q}^n_g$ $(1 \leq n \leq g)$ form an open cover of $\mathcal{Q}'_g$. By symmetricity, it is enough to show the smoothness and the geometric integrality of $\mathcal{Q}^1_g$.

Let $\mathfrak{U}_m$ be the category of $\{S; Y'_{g-2} \to ... \to Y'_0; Y_{g-1} \to ... \to Y_m\}$, where $Y'_{g-2} \to ... \to Y'_0$ is a rigid FTQ of dimension $g - 1$, and $Y_{g-1} \to ... \to Y_m$ satisfies the axioms of FTQ and

  i) the homomorphisms $E \times S \to Y_i$ induced by j_1 are monomorphisms, and there are given isomorphisms $Y_i/E \times S \cong Y'_i$ $(m \leq i \leq g - 2)$ compatible with the FTQ structures;

  ii) $\ker(Y'^{(p)}_i \to Y'^{(p)}_{i-1}) \hookrightarrow (\ker(F_{Y^{(p)}_i/S}) \cap \ker(V_{Y_i/S}))/(\alpha_p \times S)$ $(m \leq i \leq g - 2)$ under the isomorphisms in i).

Let $\mathcal{U}_m$ be the fine moduli scheme of $\mathfrak{U}_m$. Then $\mathcal{U}_0 \cong \mathcal{Q}^1_g$ and $\mathcal{U}_{g-1} \cong \mathcal{Q}'_{g-1}$. Furthermore, the truncation functor $\mathfrak{t}_m : \mathfrak{U}_{m-1} \to \mathfrak{U}_m$ induces a canonical morphism $\mathcal{U}_{m-1} \to \mathcal{U}_m$.

Given an object $\{S; Y'_{g-2} \to ... \to Y'_0; Y_{g-1} \to ... \to Y_m\}$ of $\mathfrak{U}_m$, let $G_1 = \ker(F_{Y_m/S})$. Then j_1 induces a closed immersion $\alpha_p \times S \cong G_2 \to G_1$. Let $G_3 = \ker(Y'_m \to Y'_{m-1})$. Then G_3 can be identified with a subgroup scheme of G_1/G_2 via i). Let G_4 be the inverse image of G_3 in G_1. Then we have

$$(11.1.1) \qquad G^{(p)}_4 \subset \ker(F_{Y^{(p)}_m/S}) \cap \ker(V_{Y_m/S})$$

by ii). Hence G_4 is an α-group of α-rank $m + 1$.

For convenience, in the following for a group scheme G_n such that $V_{G_n/S} = 0$, we will denote the α-sheaf of G_n by $\mathcal{F}_n$.

Locally we can choose a projection $j' : E^g \to E^m$ such that $F^{g-1-m} \circ j'$ induces a smooth homomorphism $Y_m \to E^m \times S$ which induces $G_3 \cong \alpha^m_p \times S$. By fixing a generator of the α-sheaf of α_p (note that α_p is defined over $\mathbf{F}_p$), we get generators

73

$v_0, ..., v_m$ of $\mathcal{F}_4$ such that $v_1, ..., v_m$ generate $\mathcal{F}_3$ and v_0 is a lifting of the generator $\bar{v}_0$ of $\mathcal{F}_2$. We have an exact sequence

$$(11.1.2) \qquad \alpha_p \times S \cong G_2 \hookrightarrow G_4 \twoheadrightarrow G_3 \cong \alpha_p^m \times S$$

To give a Y_{m-1} satisfying i) (for $i = m - 1$) is equivalent to giving an α-subgroup $G_5 \subset G_4$ such that the induced homomorphism $G_5 \to G_3$ is an isomorphism. This is then locally equivalent to giving a section

$$(11.1.3) \qquad v = v_0 + x_1 v_1 + ... + x_m v_m$$

of $\mathcal{F}_4$.

We give a necessary and sufficient condition on v in order that ii) is also satisfied for $i = m - 1$.

A) Let $G_6 = \ker(V_{Y_{m-1}/S})$ and $G_7 = \ker(V_{Y'_{m-1}/S})$. Then we have an exact sequence

$$(11.1.4) \qquad \alpha_p \times S \cong G_2^{(p)} \hookrightarrow G_6 \to G_7 \to 0.$$

Hence $F_{G_6/S}$ factors through G_7. Let $G_8 = \ker(Y'_{m-1} \to Y'_{m-2})$. Then $G_8^{(p)} \subset G_7$. Clearly condition ii) for $i = m - 1$ is equivalent to that the composition $G_8^{(p)} \to G_7 \to G_6^{(p)}$ is zero. This is then equivalent to that the composition of O_S-linear maps

$$(11.1.5) \qquad \mathcal{F}_6^{(p)} \xrightarrow{F_6} \mathcal{F}_7 \xrightarrow{V_8} \mathcal{F}_8^{(p)}$$

is the zero map. Locally let v'' be a section of $\mathcal{F}_6$ which maps to $\bar{v}_0^{(p)}$ in $\mathcal{F}_2^{(p)}$. Then $\mathcal{F}_6$ is generated by v'' and $\mathcal{F}_7$. Since G_8 is an α-group, the composition of $G_8^{(p)} \to G_7$ with $F_{G_7/S}$ is zero. Hence $V_8 \circ F_6(\mathcal{F}_7^{(p)}) = 0$ holds automatically. Therefore ii) for $i = m - 1$ is equivalent to $V_8 \circ F_6(v''^{(p)}) = 0$.

B) Let $Z = Y_m/G_4$ and $G_9 = \ker(V_{Z/S})$. Since $F_{Y_{m-1}/S}$ factors through Z, we have an induced homomorphism $G_6 \to G_9$. Let $G_{10} = \ker(V_{Y_m/S})$. Since $F_{Z/S}$ factors through $Y_m^{(p)}$, it induces a homomorphism $\phi : G_9 \to G_{10}^{(p)}$. Furthermore, we have an exact sequence

$$(11.1.6) \qquad 0 \to G_2^{(p)} \to G_6 \to G_9 \to G_2 \to 0$$

Hence (11.1.4) induces a closed immersion $G_7 \to G_9$ which obviously has a section. Therefore we get a sequence of O_S-linear maps

$$(11.1.7) \qquad \mathcal{F}_6^{(p)} \xrightarrow{J_6^{(p)}} \mathcal{F}_{10}^{(p)} \xrightarrow{F_{10}} \mathcal{F}_9 \xrightarrow{I_9} \mathcal{F}_7$$

whose composition is equal to F_6, and $\mathcal{F}_7$ can be identified with a subsheaf of $\mathcal{F}_9$ via the section of I_9.

Since $G_5^{(p)} \cong \ker(G_{10} \to G_6)$, we have an exact sequence

$$(11.1.8) \qquad \mathcal{F}_6 \xrightarrow{J_6} \mathcal{F}_{10} \to \mathcal{F}_5^{(p)} \to 0.$$

Furthermore, $\mathcal{F}_4^{(p)}$ can be viewed as a quotient of $\mathcal{F}_{10}$ via the closed immersion $G_4^{(p)} \hookrightarrow G_{10}$. Hence $J_6(v'')$ is a lifting of $v^{(p)}$ by (11.1.8). Conversely, any lifting of $v^{(p)}$ in $\mathcal{F}_{10}$ is equal to $J_6(v'')$ for some choice of v'' (because the only requirement on the choice of v'' is that its image in $\mathcal{F}_2$ is equal to $\bar{v}_0^{(p)}$).

Therefore condition ii) holds for $i = m-1$ iff $V_8 \circ I_9 \circ F_{10}(v'^{(p)}) = 0$ for a lifting v' of $v^{(p)}$ in $\mathcal{F}_{10}$, by A).

C) Since $V_{Z/S}$ factors through Y_m, we have an induced homomorphism $G_9 \to G_4$. One easily checks that this is an epimorphism. Hence $\mathcal{F}_4$ can be identified with a subsheaf of $\mathcal{F}_9$. Since the image of $G_6 \to G_4$ is G_5, the image of v in $\mathcal{F}_6$ (under $\mathcal{F}_9 \to \mathcal{F}_6$) is equal to 0. Hence we have $I_9(v) = 0$. Note that $F_{10}(v'^{(p)}) - v$ is a section of $\mathcal{F}_7$ because its image in $\mathcal{F}_2$ is 0. Therefore condition ii) holds for $i = m-1$ iff

$$(11.1.9) \qquad V_8(F_{10}(v'^{(p)}) - v) = 0$$

by B).

Note that $G_3^{(p)}$ can be identified with a quotient of G_{10} via $j'^{(p)}$, hence $\mathcal{F}_3^{(p)}$ is a subsheaf of $\mathcal{F}_{10}$. This also induces an epimorphism $\psi : G_9 \to G_3^{(p^2)}$ via ϕ which factors through G_7. Hence $\mathcal{F}_3^{(p^2)}$ can be identified with a subsheaf of $\mathcal{F}_7 \subset \mathcal{F}_9$.

Let $G_{11} = \ker(V_{Y'_{m-2}/S})$. Since Y'_{m-2} is a quotient of Z, we have an induced O_S-linear map $J_{11} : \mathcal{F}_{11} \to \mathcal{F}_9$. One easily checks that $\mathcal{F}_7 \subset \mathcal{F}_9$ is generated by $\mathrm{im}(J_{11})$ and $\mathcal{F}_3^{(p^2)}$, and $\mathrm{im}(J_{11}) \subset \ker(V_8)$. On the other hand, ψ induces a closed immersion $G_8^{(p)} \to G_3^{(p^2)} \cong G_3$. Note that G_8 has α-rank $m-1$. Hence $G_8^{(p)} \hookrightarrow G_3$ is defined by a nowhere-zero section

$$(11.1.10) \qquad w_1 = a_1 v_1 + \ldots + a_m v_m$$

of $\mathcal{F}_3$.

D) Fix locally a lifting v_0' in $\mathcal{F}_{10}$ of $v_0^{(p)}$. This gives a lifting of $v^{(p)}$ to

$$(11.1.11) \qquad v' = v_0' + x_1^p v_1 + \ldots + x_m^p v_m$$

in $\mathcal{F}_{10}$. Note that $w_0 = F_{10}(v_0'^{(p)}) - v_0$ is a section of $\mathcal{F}_7$. By C), we can find a section w' of $\mathcal{F}_{11}$ such that

$$(11.1.12) \qquad w_0 = J_{11}(w') - b_1 v_1 - \ldots - b_m v_m$$

for some sections $b_1, \ldots, b_m$ of O_S. We have

$$(11.1.13)
\begin{aligned}
F_{10}(v'^{(p)}) - v &= F_{10}(v_0'^{(p)}) - v_0 + \sum_{i=1}^m (x_i^{p^2} - x_i) v_i \\
&= w_0 + \sum_{i=1}^m (x_i^{p^2} - x_i) v_i \\
&= J_{11}(w') + \sum_{i=1}^m (x_i^{p^2} - x_i - b_i) v_i.
\end{aligned}$$

75

Hence $w = F_{10}(v'^{(p)}) - v - J_{11}(w')$ is a section of $\mathcal{F}_3$.

By C), condition ii) holds for $i = m - 1$ iff $V_8(w) = 0$. This is then equivalent to that $w = tw_1$ (see (11.1.10)) for some section t of O_S, or explicitly

$$(11.1.14) \qquad x_i^{p^2} - x_i = a_i t + b_i \quad (1 \le i \le m).$$

Now we see that $\mathcal{U}_{m-1} \to \mathcal{U}_m$ is locally defined by variables $t, x_1, ..., x_m$ with defining equations (11.1.14) for some a_i, b_i $(1 \le i \le m)$ in $O_{\mathcal{U}_m}$, hence it is smooth. One checks easily that $a_2/a_1, ..., a_m/a_1$ are algebraically independent over $\mathbf{F}_p$. Hence by Corollary 7.8, we see that $\mathcal{U}_{m-1}$ is geometrically integral if $\mathcal{U}_m$ is so.

Proof of ii). For any supersingular abelian variety X of dimension g over k, we have a minimal isogeny $\rho_0 : E^g \otimes k \to X$. Choose an arbitrary factor E of E^g. Then the induced homomorphism $E_1 := E \otimes k \to X$ is a closed immersion (for otherwise we get a contradiction because ρ_0 is minimal). By induction on g, we may assume $X' := X/E_1$ is the end of a rigid FTQ, i.e. there is an isogeny $\rho_1 : E^{g-1} \otimes k \to X'$ inducing a rigid FTQ structure. Then ρ_1 can be lifted to an isogeny $\rho_2 : Y \to X$, where Y is an extension of $E^{g-1} \otimes k$ by E_1. Let $G = \ker(V_{Y/k})$ and $f : \alpha_p \hookrightarrow Y^{(p)}$ be induced by $E_1^{(p)} \hookrightarrow Y^{(p)}$. Then $G' := G/f(\alpha_p) \cong \alpha_p^{g-1}$. Let $Y_{g-1} = Y^{(p)}/f(\alpha_p)$. Then $\rho_2 \circ V_{Y/k} : Y^{(p)} \to X$ factors through Y_{g-1}, and $G' \hookrightarrow Y_{g-1}$. On the other hand $E_1 \hookrightarrow Y_{g-1}$ induces $\alpha_p \hookrightarrow Y_{g-1}$. Therefore Y_{g-1} has an α-subgroup $\alpha_p \times G'$ of α-rank g, i.e. $a(Y_{g-1}) = g$. Hence Y_{g-1} is superspecial by Fact 1.6. Finally, it is easy to see that $Y_{g-1} \to X$ induces a rigid FTQ structure.

Proof of iii). Let $\mathfrak{V}_g$ be the category of isogenies $f : E^g \times S \to Y$ such that $\ker(f)$ is an α-group of α-rank $g - 1$. Then the fine moduli scheme of $\mathfrak{V}_g$ is isomorphic to $\mathbf{P}^{g-1}$. The truncation functor from $\mathfrak{Q}'_g$ to $\mathfrak{V}_g$ defined by

$$(11.1.15) \qquad \{Y_{g-1} \to ... \to Y_0\} \mapsto \{Y_{g-1} \to Y_{g-2}\}$$

induces a canonical morphism $\phi : \mathcal{Q}'_g \to \mathbf{P}^{g-1}$. (Note that ϕ is an epimorphism by Proposition 10.1.i).) Let D_0 be the union of hyperplanes

$$(11.1.16) \qquad c_1 x_1 + c_2 x_2 + ... + c_g x_g = 0 \quad (c_1, c_2, ..., c_g \in \mathbf{F}_{p^2}, \text{ not all } = 0)$$

in $\mathbf{P}^{g-1}$. Then $\phi^{-1}(D_0)$ is a divisor $D \subset \mathcal{Q}'_g$. Clearly $a((Y_0)_s) > 1$ iff $s \in D$ by Fact 5.8.

11.2. Proof of Corollary 10.3.

For any given polarization η of $E^g \otimes \bar{\mathbf{F}}_p$, such that $\ker(\eta) = \ker(F^{g-1})$, the FTQ structure of a rigid PFTQ with respect to η uniquely determines its PFTQ structure. Hence the canonical morphism $\mathcal{P}'_{g,\eta} \to \mathcal{Q}'_g$ is a closed immersion. From Proposition 10.1.iii) we see that

$$(11.2.1) \qquad D_\eta := \mathcal{P}'_{g,\eta} \cap D$$

is a divisor in $\mathcal{P}'_{g,\eta}$. Note that the morphism Ψ in (4.2.1) is quasi-finite and

$$(11.2.2) \qquad S_{g,1}(a \geq 2) \otimes \bar{F}_p = \Psi(\coprod_{\eta \in \Lambda} D_\eta).$$

Hence every irreducible component of $\mathcal{S}_{g,1}(a \geq 2)$ is of codimension 1 in $\mathcal{S}_{g,1}$.
Q.E.D.

11.3. Sketch of a proof of Proposition 4.3.

Again we will use double induction. What we will do is translating the argument of Chapter 7 to a direct argument on group schemes.

We use the same notation as that in 3.9. Let $K = \mathbf{F}_{p^2}$ and $G_\alpha = (E \otimes K)[F^{g-1}]$. Let $f : E \otimes K \to E^g \otimes K$ be a homomorphism such that

$$(11.3.1) \qquad \ker(f^t \circ \eta \circ f) \supset (E \otimes K)[F^g]$$

(there are many such homomorphisms f, see Remark 6.1). Thus f induces a homomorphism $\Theta : G_\alpha \hookrightarrow G$ such that

$$(11.3.2) \qquad \Theta^D \circ \theta \circ \Theta = 0$$

(for θ see 3.9, note that Θ^D is an epimorphism). Let

$$(11.3.3) \qquad G' = \ker(\Theta^D)/\mathrm{im}(\Theta).$$

Then θ induces an isomorphism $\theta' : G' \to G'^D$.

Let $\mathfrak{U}_m$ be the category of pairs of filtrations $\{G'_{g-3} \subset ... \subset G'_0; G_{g-1} \subset ... \subset G_m\}$ of flat group schemes over some K-scheme S such that

a) $\{G'_{g-3} \subset ... \subset G'_0\}$ is a rigid PFTQ of group schemes of genus $g-2$ (with respect to θ');

b) $G_m, ..., G_{g-1}$ are closed subgroup schemes of $G \times S$ satisfying
   b1) $G_{g-1} = 0$;
   b2) G_{i-1}/G_i is an α-group of α-rank i ($g < i < m$);
   b3) $f_i^D \circ \theta_S \circ f_i = 0$, where $f_i : G_i \to G \times S$ is the inclusion and $\theta_S = \theta \times \mathrm{id}_S$, and $F^j \circ V^{i-j} = 0$ on $\ker(f_i^D \circ \theta_S)/G_i$ ($0 \leq j \leq i/2, m \leq i < g$);
   b4) $G_i = G_m \cap (G[F^{g-1-i}] \times S)$ ($m < i < g$);

c) the composition of the projection $G \to G/G_m$ with $\Theta_S = \Theta \times \mathrm{id}_S$ is a closed immersion, and Θ_S induces isomorphisms

$$(11.3.4) \qquad \ker(\Theta_S^D \circ \theta_S \circ f_i)/(\mathrm{im}(\Theta_S) \cap G_i) \cong G'_{i-1} \quad (\max(m,1) \leq i \leq g-2)$$

compatible with the filtrations;

d) $(G'_{i-2}/G'_{i-1}) \subset ((G/G_i)^{(p)}[F] \cap (G/G_i)^{(p)}[V])/\Theta_S(G_\alpha[F] \times S)$ ($\max(m,2) \leq i \leq g-2$) under the isomorphisms of c), where $(G/G_i)^{(p)}[V] = \ker(V : (G/G_i)^{(p)} \to (G/G_i))$.

Let $\mathcal{U}_m^\Theta$ be the fine moduli scheme of $\mathfrak{U}_m^\Theta$. Then $\{\mathcal{U}_0^\Theta | \forall \Theta\}$ is an open cover of $\mathcal{P}'_g$ and $\mathcal{U}_{g-1}^\Theta \cong \mathcal{P}'_{g-2}$. Furthermore, the truncation functor $\mathfrak{t}_m : \mathfrak{U}_{m-1}^\Theta \to \mathfrak{U}_m^\Theta$ induces a canonical morphism $\mathcal{U}_{m-1}^{\Theta} \to \mathcal{U}_m^{\Theta}$. Our task is to show this morphism is smooth of relative dimension 1.

Suppose we are given an object $\{G'_{g-3} \subset \ldots \subset G'_0; G_{g-1} \subset \ldots \subset G_m\}$ of $\mathfrak{U}_m^{\Theta}$ over some S, where $m \geq 2$. Let $G_\beta = \ker(f_m^D \circ \theta_S)/G_m$. Then θ induces $\theta_1 : G_\beta \cong G_\beta^D$. Let $G_\gamma = G_\beta[F]$. One checks that G_γ is flat over S (it is enough to check that $\omega_{G_\gamma/S}$ is flat).

Let $C.$ be the complex

(11.3.5)
$$G_\alpha \times S \overset{\Theta_S}{\hookrightarrow} G_\beta \overset{\Theta_S^D \circ \theta}{\longrightarrow} G_\alpha^D[F^m] \times S.$$

Then the dual of $C.$ induces a complex

(11.3.6)
$$C_.^{D1} : \alpha_p \times S \hookrightarrow G_\gamma^D \twoheadrightarrow \alpha_p \times S.$$

Let $r = [m/2]$ and $s = m - 2r$. By condition b3), on G_β the homomorphism $p^r F^s$ equals 0, hence induces

(11.3.7)
$$G_m^D \to G_\beta^{(p^s)}.$$

Since $G_\gamma \cong \operatorname{coker}(V_{G_\beta^D/S})$, (11.3.7) induces $G_\gamma \to G_\beta^{(p^{s+1})}$, hence induces a morphism of complexes

(11.3.8)
$$\Phi_. : C_.^{D1} \to C_.^{(p^{s+1})}.$$

One checks that Φ_0 and Φ_2 are closed immersions but $H_1(\Phi.) = 0$.

Therefore we get a flat quotient G_δ ($\subset \ker(V_{G_\beta^{(p^s)}/S})$) of G_γ^D. By condition b3), we have $p^{r-1}F^{s+2} = 0$ on G_β also, hence $F_{G_\delta/S} = 0$. Therefore G_δ is an α-group of α-rank 2. Hence G_δ^D is a closed subgroup scheme of G_γ.

On the other hand, $C.$ induces a complex

(11.3.9)
$$C_.^1 : G_\epsilon = \alpha_p \times S \hookrightarrow G_\gamma \twoheadrightarrow \alpha_p \times S.$$

By condition c), this induces a closed immersion

(11.3.10)
$$G'_\gamma := G'_{m-2}/G'_{m-1} \hookrightarrow H_1(C_.^1).$$

Let G_ζ be the inverse image of G'_γ in G_γ. By condition d), we have $V_{G_\zeta/S} = 0$, hence G_ζ is an α-group.

Note that $G_\epsilon \subset G_\delta^D$. Hence G_δ^D and G_ζ generate a closed subgroup scheme $G_\lambda \subset G_\gamma$ which is an α-group of α-rank $m + 1$.

Let $\mathcal{F}$ be the α-sheaf of G_λ. Locally one can choose an O_S-basis $s_0, s_g, s_1, \ldots,$ s_{m-1} of $\mathcal{F}$, where s_0 comes from G_ϵ, s_g comes from G_δ/G_ϵ and $s_1, \ldots, s_{m-1}$ come from G'_γ. (We can choose $s_0, s_g, s_1, \ldots, s_{m-1}$ in a "standard" way, i.e. choose generators of the α-sheaf of some $(E^{m+1} \otimes K)[F]$, as in the argument in 11.1.)

We check the following facts one by one:

78

i) To give a flat subgroup scheme $G_{m-1} \subset G \times S$ containing G_m satisfying b4), c) (for $i = m-1$) and b2) (for $i = m$) is equivalent to giving a flat subgroup scheme of G_λ of α-rank m. This is locally equivalent to giving a section

$$(11.3.11) \qquad s = s_0 + x s_g + x_1 s_1 + \ldots + x_{m-1} s_{m-1}$$

of $\mathcal{F}$.

ii) In order that G_{m-1} satisfies b3) (for $i = m$) and d) (for $i = m-1$, when $m > 2$) also, it is equivalent to that the "translated version of condition 7.9.A)" hold (see Lemma 7.9 and Lemma 7.10).

For condition d), we use the same argument as that in 11.1. Then we get conditions on s, of the form (11.1.14).

iii) For condition b3), we first note that (11.3.7) induces a homomorphism of α-groups

$$(11.3.12) \qquad G_\lambda^D \to G_\lambda^{(p^{s+1})}$$

which is equivalent to an O_S-bilinear form

$$(11.3.13) \qquad \langle , \rangle : \mathcal{F} \otimes_{O_S} \mathcal{F}^{(p^{s+1})} \to O_S.$$

Then we can show that condition b3) is equivalent to the "translated version of condition 7.10.B')", i.e. $\langle s, s^{(p^{s+1})} \rangle = 0$.

iv) Writing down the explicit equations of ii) and iii), we get exactly (7.11.1), where t is a free variable and a_i, b_i, c_i, d, e are pull-backs of sections in $O_{\mathcal{U}_m^\Theta}$.

v) Therefore the morphism $\mathcal{U}_{m-1}^\Theta \to \mathcal{U}_m^\Theta$ induced by t_m is locally defined by variables $x, x_1, \ldots, x_{m-1}, t$ satisfying defining relation (7.11.1), hence is smooth of relative dimension 1.

The remaining arguments are the same as that of Lemma 7.14.

11.4. Proof of Proposition 10.2.

Note that i) is clear by Fact 2.2. For ii), we only need to note that a rigid FTQ $\{Y_{g-1} \to \ldots \to Y_0\}$ is uniquely determined by $Y_{g-1} \to Y_0$. Finally, iii) is clear by the argument of Remark 6.4.

11.5. The reducibility of $\mathcal{D}_g$ for $g > 2$.

Remark. When $g > 2$, the morphism $\mathcal{Q}_g \to \mathcal{D}_g$ is not an epimorphism. Indeed, if $g > 2$, we can construct an isogeny $\rho : E^g \to Y \cong E^g$ of degree $p^{g(g-1)/2}$ such that $\ker(\rho) \cong \ker(F_{E/k}^{g(g-1)/2})$, which clearly can not be extended to an FTQ. Since $\mathcal{Q}_g$ is proper, this also shows that $\mathcal{D}_g$ is not irreducible when $g > 2$.

For any polarization η of E^g and any $d > 0$, let $\mathfrak{B}_{g,d,\eta}$ be the category of pairs $\{\rho, \eta_0\}$, where $\rho : E^g \times S \to Y$ is an isogeny of degree $p^{g(g-1)/2}$ and η_0 is a polarization of Y of degree d^2 compatible with η. Then $\mathfrak{B}_{g,d,\eta}$ has a fine moduli scheme $\mathcal{B}_{g,d,\eta}$. When $d = 1$ and $\ker(\eta) = \ker(F^{g-1})$, the forgetful functor $\mathfrak{B}_{g,\eta} \to \mathfrak{B}_{g,1,\eta}$ gives a canonical morphism $\mathcal{P}_{g,\eta} \to \mathcal{B}_{g,1,\eta}$ whose restriction on $\mathcal{P}'_{g,\eta}$ is an isomorphism. Furthermore, when $g > 3$, $\mathcal{P}_{g,\eta} \to \mathcal{B}_{g,1,\eta}$ is not surjective (for example, when $g = 4$, we can construct $\rho : E^g \to Y$ and η_0 such that $\ker(\rho) \cong \ker(F^3_{E^2/k})$). Hence $\mathcal{B}_{g,1,\eta}$ is not irreducible when $g > 3$.

11.6. A lemma on polynomials.

For the proof of Proposition 10.4.i), ii) we need two lemmas. The first is the following:

Lemma. Let $x_1, ..., x_g$ be free variables over $K \supset \mathbf{F}_{p^2}$. Let y_i $(1 \leq i \leq g)$ be the (g,i)-minor of the matrix

(11.6.1)
$$\begin{pmatrix} x_1 & \cdots & x_g \\ \vdots & \ddots & \vdots \\ x_1^{p^{2g-2}} & \cdots & x_g^{p^{2g-2}} \end{pmatrix}$$

Then

     i) $y_1, ..., y_g$ are algebraically independent over K;

     ii) for any non-zero $g \times g$ matrix (c_{ij}) over $\mathbf{F}_{p^2}$ and any $n > 0$,

(11.6.2)
$$\begin{vmatrix} x_1 & \cdots & x_g \\ \vdots & \ddots & \vdots \\ x_1^{p^{2g-4}} & \cdots & x_g^{p^{2g-4}} \\ \sum_j c_{1j} y_j^{p^n} & \cdots & \sum_j c_{gj} y_j^{p^n} \end{vmatrix} \neq 0.$$

Proof. First, it is trivial to see that $y_1, ..., y_g$ are linearly independent over $\mathbf{F}_{p^2}$. Then by Fact 5.8, the vectors

(11.6.3)
$$(y_1^{p^{2i}}, ..., y_g^{p^{2i}}) \quad (0 \leq i \leq g-1)$$

are linearly independent over $K(x_1, ..., x_g)$. It is easy to check that

(11.6.4)
$$x_1^{p^{2g-4}} y_1^{p^{2i}} + ... + x_g^{p^{2g-4}} y_g^{p^{2i}} = 0 \quad (0 \leq i \leq g-2)$$

On the other hand, if z_i $(1 \leq i \leq g)$ are the (g,i)-minors of the (non-singular) matrix

(11.6.5)
$$\begin{pmatrix} y_1 & \cdots & y_g \\ \vdots & \ddots & \vdots \\ y_1^{p^{2g-2}} & \cdots & y_g^{p^{2g-2}} \end{pmatrix}$$

then obviously we have

(11.6.6)
$$z_1 y_1^{p^{2i}} + ... + z_g y_g^{p^{2i}} = 0 \quad (0 \leq i \leq g-2).$$

80

Therefore we have

$$(11.6.7) \qquad x_i^{p^{2g-4}} / x_1^{p^{2g-4}} = z_i/z_1 \quad (1 < i \le g).$$

Hence x_i/x_1 ($1 < i \le g$) are algebraic over $K(y_1, ..., y_g)$. Furthermore, using Fact 5.8, it is easy to show that x_1 is algebraic over $K(y_1, ... y_g, x_2/x_1, ..., x_g/x_1)$. Therefore i) is proved.

For ii), note that the left hand side of (11.6.2) is equal to $(-1)^{g-1} \sum_{i,j} c_{ij} y_i y_j^{p^n}$, which is obviously non-zero by i).   Q.E.D.

11.7. A lemma on Dieudonné modules.

The second lemma we need is the following:

Lemma. *Let M be a supergeneral Dieudonné module of genus $g > 2$ over $W = W(k)$ and $f : M^t \to M$ be an A-monomorphism. If M is general enough (see 10.2), then $f(M^t) \subset F^{g-2} S(M)$.*

Proof. Since M is supergeneral, so is M^t. Choose a set of generators of the skeleton of $S(M)$ and let $(a_1, ..., a_g)$ be a generator of $M/(F, V)M \cong k$. Note that $M \cap F^m S(M) = (F, V)^m M$. Hence for any $m < g - 1$, the subspace

$$(11.7.1) \qquad (M \cap F^m S(M))/M \cap F^{m+1} S(M) \subset F^m S(M)/F^{m+1} S(M) \cong k^{\oplus g}$$

is generated by

$$(11.7.2) \qquad (a_1^{p^{m-2i}}, ..., a_g^{p^{m-2i}}) \quad (0 \le i \le m)$$

and the vectors in (11.7.2) are linearly independent. We can choose generators of the skeleton of M^t such that any generator $(b_1, ..., b_g)$ of $M^t/(F, V)M^t$ satisfies

$$(11.7.3) \qquad \sum_i b_i a_i^{p^{g-2j}} = 0 \quad (1 \le j \le g - 1)$$

Hence we may choose $b_1, ..., b_g$ such that $d_i = b_i^{p^{g-2}}$ ($1 \le i \le g$) is equal to the (g, i)-minor of

$$(11.7.4) \qquad \begin{pmatrix} a_1 & \cdots & a_g \\ \vdots & \ddots & \vdots \\ a_1^{p^{2g-2}} & \cdots & a_g^{p^{2g-2}} \end{pmatrix}$$

Suppose that $f(M^t) \subset F^m S(M)$ but $f(M^t) \not\subset F^{m+1} S(M)$ ($0 \le m < g - 2$). We now show that $a_1, ..., a_g$ satisfy a non-trivial homogeneous relation. Note that f maps the skeleton of M^t into the skeleton of M. Hence f induces a linear map

$$(11.7.5) \qquad \bar{f} : k^{\oplus g} \cong S(M^t)/FS(M^t) \to F^m S(M)/F^{m+1} S(M) \cong k^{\oplus g}$$

81

which corresponds to a (non-zero) $g \times g$ matrix (c_{ij}) over $\mathbb{F}_{p^2}$. Therefore

(11.7.6) $\qquad (\sum_j c_{1j} b_j, ..., \sum_j c_{gj} b_j) \in (M \cap F^{-1} S_0(M))/S_0(M)$

is linearly dependent with the vectors in (11.7.2). Hence

(11.7.7) $\qquad \begin{vmatrix} a_1 & \cdots & a_g \\ \vdots & \ddots & \vdots \\ a_1^{p^{2g-4}} & \cdots & a_g^{p^{2g-4}} \\ \sum_j c_{1j} d_j^{p^{g-2-m}} & \cdots & \sum_j c_{gj} d_j^{p^{g-2-m}} \end{vmatrix} = 0.$

This is a non-trivial condition on $a_1, ..., a_g$ when $m < g-2$ by Lemma 11.6.   Q.E.D.

11.8. Quasi-polarizations of a supergeneral Dieudonné module which is general enough.

Corollary. If a supergeneral Dieudonné module M of genus g is general enough, then $S(M)^t \subset F^{2g-3} S(M)$ for any quasi-polarization of M.

11.9. Induced quasi-polarizations on Dieudonné submodules.

Proposition. Let M be a supergeneral Dieudonné module of genus g over $W = W(k)$. Then any quasi-polarization of $S(M)$ such that $S(M)^t \subset F^{2g-3} S(M)$ induces a quasi-polarization on M.

Proof. In this case we have $S(M^t) \subset F^{g-2} S(M)$. This induces a linear map

(11.9.1) $\qquad f : k^{\oplus g} \cong P := S(M^t)/FS(M^t) \rightarrow F^{g-2} S(M)/F^{g-1} S(M) \cong k^{\oplus g}$

Note that

(11.9.2) $\quad F^{g-2} S(M)/F^{g-1} S(M) \cong F^{-1} S_0(M)/S_0(M) \cong (S(M^t)/FS(M^t))^t = P^t$

We see that f is equivalent to an alternating form on P. Let v be a generator of $M^t/(F, V)M^t \cong k$. Then canonically

(11.9.3) $\qquad Q := F^{g-2} S(M) \cap M/F^{g-1} S(M) \cong \ker(P^t \rightarrow (kv)^t).$

By $\langle v, v \rangle = 0$ we have $f(v) \in Q$. This shows that $M^t \subset M$, which is just what we need to prove.   Q.E.D.

11.10. The colength of quasi-polarizations of a supergeneral Dieudonné module which is general enough.

Remark. When g is even, we have a quasi-polarization of $N = A_{1,1}^{\oplus g}$ such that $N^t = F^{2g-3} N$. However, this is impossible when g is odd because the colength of

N^t in N is an even number in any case (one can also see this by noting that the alternating form on P in (11.9.1) is degenerate when g is odd). Therefore when M is general enough, the colength of M^t in M is at least $g(g-2)$ when g is even, and $g(g-2)+1$ when g is odd. (cf. Remark 10.4 also.)

11.11. Another fact about the colength of quasi-polarizations of a super-general Dieudonné module.

For the proof of Proposition 10.4.iii) we need the following lemma.

Lemma. Let $n < g/2$ be a positive integer $(g > 2)$. Then there is no supergeneral quasi-polarized Dieudonné module M of genus g over $W = W(k)$ such that the colength of M^t in M equals $2n$.

Proof. We prove this by contradiction. Assume that $a(M) = 1$ and the colength of M^t in M is $2n$. Then $S_0(M^t) = (S^0 M)^t$ has colength $2n$ in $S_0 M = F^{g-1} S^0 M$. By Proposition 6.1, we can factor $S^0 M$ to a direct sum of quasi-polarized Dieudonné modules N and N' such that $N^t \subset F^g N$ and $N'^t = F^{g-1} N'$, where $N' \neq 0$ because $2n < g$. We have $F^{g-2} S^0(M^t) \subset F^{g-1} N \oplus F^{g-2} N'$. Let $M' = M^t \cap F^{g-2} S^0(M^t)$. Then M' has colength 1 in $F^{g-2} S^0(M^t)$. On the other hand, $F^{g-2} S^0(M^t) \cap N \not\subset M'$. Hence $(F^{g-2} S^0(M^t) \cap N) + M' = F^{g-2} S^0(M^t) \subset M$. This is impossible since $S_0 M = F^{g-1} N \oplus F^{g-1} N'$ is the largest superspecial A-submodule of M but $F^{g-2} S^0(M^t) \not\subset S_0 M$.   Q.E.D.

11.12. Proof of Proposition 10.4.

Note that Proposition 10.4.i) is equivalent to Corollary 11.8, Proposition 10.4.ii) is immediate by Proposition 11.9, and Proposition 10.4.iii) is immediate by Lemma 11.11.

11.13. A generic point of $\mathcal{S}_{g,d}$ may not represent a supergeneral abelian variety.

Note that Proposition 10.4.iii) gives:

Corollary. Let $g > 2$. If $p^n \| d$ for some $0 < n < g/2$, then any generic point of $\mathcal{S}_{g,d}$ does not represent a supergeneral abelian variety (i.e. $a(-/V) > 1$ for any $V \subset \mathcal{S}_{g,d}$).

This phenomenon has already been observed in [36, Remark 6.10.c)] for $g = 3$.

12. Examples on $\mathcal{S}_{g,d}$ $(d{>}1)$

In this chapter we describe the structure of $\mathcal{S}_{g,d}$ for some low values of g and $d > 1$.

12.1. Example, $g = 1$.

The number of geometric points of $\mathcal{S}_{1,d}$ is equal to the class number h_p of B (see (9.1.2)), in particular it is independent of d. In fact, for any supersingular elliptic curve E_0 over k and any polarization μ of E_0 of degree d^2, we have $d|\mu$. Hence $\mu \leftrightarrow \mu/d$ gives a one-to-one correspondence

{supersingular elliptic curves over k together with a polarization of degree d^2}

$\longleftrightarrow$ {principally polarized supersingular elliptic curves over k}

which induces a one-to-one correspondence

$$(12.1.1) \qquad\qquad \mathcal{S}_{1,d} \otimes k \longleftrightarrow \mathcal{S}_{1,1} \otimes k.$$

The situation is quite different for $g > 1$. For example, when $p \nmid d$, to give a polarization of $E^g \otimes k$ of degree d^2, one first needs to choose a sequence of "elementary divisors" $\delta = (\delta_1, ..., \delta_g)$ of positive integers $\delta_1|\delta_2...|\delta_g$ such that $\delta_1 \cdots \delta_g = d$ (see [53, p. 458] and [33, Chapter 1]). From this we see that the number of superspecial points and the number of irreducible components in $\mathcal{S}_{g,d}$ both depend on d.

12.2. General procedure of calculating the components of $\mathcal{S}_{g,d}$.

For simplicity, in the following we consider the case when d is a power of p. The method to calculate the components of $\mathcal{S}_{g,d}$ is roughly as follows.

Step 1. We introduce some sets of integer invariants. For each combination we check if it is possible (i.e., if there is a polarized supersingular abelian variety having all the integer invariants in the combination). If it is possible, then the corresponding polarized abelian varieties together with a minimal isogeny have a fine moduli scheme T whose canonical image in $\mathcal{S}_{g,d}$ is in fact a locally closed subset.

Step 2. It is an algebraic problem to calculate for such a fine moduli scheme T its dimension and the number of irreducible components.

Step 3. After getting enough information about T, we check the two possibilities: either the images of each irreducible component of T in $\mathcal{S}_{g,d}$ dominates an irreducible component of $\mathcal{S}_{g,d}$, in which case we use the corresponding combination of integer invariants to index the closure of the image of T in $\mathcal{S}_{g,d}$ (which usually consists of irreducible components of the same dimension whose number is equal to a class number or a sum of class numbers); or the image of T in $\mathcal{S}_{g,d}$ does not dominate any irreducible component of $\mathcal{S}_{g,d}$ (i.e., such kind of polarized abelian varieties can be deformed to some more general case), in which case we delete the corresponding combination of integer invariants from our calculation table.

For simplicity we do calculation on Dieudonné modules. In addition to the dimension g, the degree d^2 of the polarization and the a-number a we will introduce the following integer invariants.

a) $s = (s_1, s_2, ...)$, where

$$(12.2.1) \qquad s_i = \dim_k(M \cap F^{i-1} S^0 M)/(M \cap F^i S^0 M) \ (1 \le i \le g - 1).$$

It is known (see [45, p.337]) that $s_i < s_{i+1}$ unless $s_i = g$. For simplicity we omit those $s_i = g$.

Note that s depends only on the abelian variety but not on the polarization, and $s = (1, 2, ..., g - 1)$ when $a = 1$.

b) $r = (r_1, ..., r_g)$: by Proposition 6.1 we can decompose $S^0 M = Ax_1 \oplus ... \oplus Ax_g$ (where $x_1, ..., x_g$ are in the skeleton of $S^0 M$) such that

$$(12.2.2) \qquad (S^0 M)^t = S_0(M^t) = AF^{r_1} x_1 \oplus ... \oplus AF^{r_g} x_g \ (r_1 \le ... \le r_g).$$

Note that the odd numbers in r can only appear in pairs.

For Step 2, we don't have a tool like PFTQ. Instead we often consider the possible structure of M^t first: after fixing the integer invariants, the only requirement on M^t is $\langle M^t, M^t \rangle \subset W$.

For Step 3, the following observation is helpful: after deformation, the dimension of the corresponding T will increase, the lexicographic order of s will decrease, at least one $r_i \in r$ will increase while the others in r not decrease, and a will not increase. Proposition 10.4.iii) is also helpful.

12.3. Example, $g = 2$.

We know (see Remark 10.6) that every irreducible component of $\mathcal{S}_{2,d}$ is of dimension 1, and $a(-/V) = 1$ for each irreducible component $V \subset \mathcal{S}_{2,d}$.

If $d = p$, then r can only be $(2,2)$, and the number of irreducible components of $\mathcal{S}_{2,p}$ is equal to $H_2(p, 1)$.

If $d = p^2$, then r can be either $(3,3)$ or $(2,4)$. The number of irreducible components with $r = (3, 3)$ in $\mathcal{S}_{2,p^2}$ is equal to $H_2(1, p)$ and the number of irreducible components with $r = (2, 4)$ is equal to $\#(\mathcal{L}_{\mathfrak{f}, \sim}/\sim)$, where $\mathfrak{f} = \{f_l \in M_2(\mathcal{O}_l)\}$, $f_l = I_2$ for each $l \ne p$ and $f_p = \mathrm{diag}(1, p)$ (see 8.5).

12.4. Example, $g = 3$.

The cases are already discussed in [36, Remark 6.10].

If $d = p$, then by Proposition 10.4.iii) we have $a(-/V) \ge 2$ for any irreducible component $V \subset \mathcal{S}_{3,d}$. One checks that either $r = (2, 2, 2)$, $s = (1)$ or $r = (1, 1, 2)$, $s = (2)$ (both with $a = 2$). Every irreducible component of $\mathcal{S}_{3,p}$ has dimension 2. The number of irreducible components with $s = (1)$ is equal to $H_3(p, 1)$, and the number of irreducible components with $s = (2)$ is equal to $H_3(1, p)$ by Corollary 4.8.ii).

Hence the total number of irreducible components of $\mathcal{S}_{3,p}$ is equal to $H_3(p,1) + H_3(1,p)$.

If $d = p^2$, then Theorem 10.5.ii) says that $\mathcal{S}_{3,p^2}$ has $H_3(1,p)$ irreducible components of dimension 3 with $a = 1$ and $r = (3,3,4)$. One checks that these are all of the irreducible components.

12.5. Example, $g = 4$.

The calculation is tedious and we only list the results below. In the following table "a" means $a(-/V)$, "dim" means the dimension of each irreducible component V with the given invariants d, a, s, r, "# " means the number of such irreducible components, and each C_i is a class number. We have $C_i = \#(\mathcal{L}_{f^i,\sim}/\sim)$ $(1 \le i \le 3)$, where $f^i = \{f_l^i \in M_4(\mathcal{O}_l)\}$, and

i) $f_l^i = I_4$ for each i and each $l \ne p$;

ii) $f_p^1 = \mathrm{diag}(1,1,1,p)$, $f_p^2 = \begin{pmatrix} 0 & F & 0 & 0 \\ -F & 0 & 0 & 0 \\ 0 & 0 & 1 & 0 \\ 0 & 0 & 0 & 1 \end{pmatrix}$, $f_p^3 = \mathrm{diag}(1,1,1,p^2)$.

Table of irreducible components of $\mathcal{S}_{4,d}$ $(d|p^4)$

d	a	s	r	dim	#
1	1	$(1,2,3)$	$(3,3,3,3)$	4	$H_4(1,p)$
p	2	$(1,2)$	$(3,3,3,3)$	4	$H_4(1,p)$
p	2	$(2,3)$	$(2,2,2,2)$	4	$H_4(p,1)$
p^2	1	$(1,2,3)$	$(4,4,4,4)$	5	$H_4(p,1)$
p^2	2	$(1,3)$	$(3,3,3,3)$	5	$H_4(1,p)$
p^3	1	$(1,2,3)$	$(4,4,4,6)$	5	C_1
p^3	2	$(1,2)$	$(4,4,4,4)$	5	$H_4(p,1)$
p^3	2	$(1,3)$	$(3,3,4,4)$	5	C_2
p^3	2	$(2,3)$	$(3,3,3,3)$	5	$H_4(1,p)$
p^4	1	$(1,2,3)$	$(5,5,5,5)$	6	$H_4(1,p)$
p^4	1	$(1,2,3)$	$(4,4,4,8)$	5	C_3
p^4	2	$(1,2)$	$(4,4,4,6)$	5	C_1

13. A scheme-theoretic definition of supersingularity

In 1.10 we defined the supersingular locus in $\mathcal{A}_{g,1,n} \otimes \mathsf{F}_p$ ($n > 2, p \nmid n$) to be the subset $\mathcal{S}_{g,1,n} \subset \mathcal{A}_{g,1,n} \otimes \mathsf{F}_p$ of supersingular points with the reduced induced scheme structure, knowing that $\mathcal{S}_{g,1,n}$ is an algebraic subset of $\mathcal{A}_{g,1,n}$ (cf. [38, Corollary 2.3.2] or [59, Theorem 1.2]). However, in this way we can not define supersingular abelian schemes over a general (especially non-reduced) base, therefore we can hardly define $\mathcal{S}_{g,1,n}$ as a fine moduli scheme (see Remark 13.15 below).

In this chapter we give a definition of "supersingular abelian schemes", and then give a *new* definition of the supersingular locus as a fine moduli scheme, i.e. the scheme representing the functor from ((F$_p$-schemes)) to ((sets)) of supersingular abelian schemes.

13.1. Generic flatness.

The terminologies in the following two definitions are not commonly used, and we take them for technical use in this chapter only.

Definition. Let $\mathcal{F}$ be a coherent sheaf over a noetherian scheme S. We say $\mathcal{F}$ is *generically flat* (of rank $\geq r$) over S if for any associated point $\xi \in S$ with (a choice of) primary ideal $Q_\xi \subset O_{S,\xi}$, the stalk $\mathcal{F}_\xi$ of $\mathcal{F}$ at ξ satisfies that $\mathcal{F}_\xi / Q_\xi \mathcal{F}_\xi$ is free (of rank $\geq r$) over $O_{S,\xi}/Q_\xi$.

We say a finite group scheme $\pi : G \to S$ is *generically flat* of rank (resp. p-rank) $\geq r$ if $\pi_* O_G$ is generically flat of rank $\geq r$ (resp. p^s for some $s \geq r$) over S.

13.2. Epimorphism.

Definition. A morphism of noetherian schemes $T \to S$ is called an *epimorphism* if it is set-theoretically surjective and for any associated point $\xi \in S$ with primary ideal $Q_\xi \subset O_{S,\xi}$, the pull-back $T \times_S \operatorname{Spec}(O_{S,\xi}/Q_\xi)$ is flat over $\operatorname{Spec}(O_{S,\xi}/Q_\xi)$.

13.3. Strictly supersingular abelian schemes.

Next let S be a noetherian F_p-scheme.

Definition. An abelian scheme X over S of dimension g is called *strictly supersingular* if

i) for any choice of non-negative integers $n < g$, the two finite group schemes $\bigcap_{0 \leq i \leq n} X[p^{n-i}F^{2i}]$ and $X[p^{g-1}] \cap \bigcap_{1 \leq i \leq g-1} X[p^{g-1-i}F^{2i+n}]$) are generically flat;

ii) the p-rank of $\bigcap_{0 \leq i \leq g(g-1)/2+1} X[p^{g(g-1)/2+1-i}F^{2i}]$ at any generic point is at least $g(2g^2 - 3g + 5)/2$.

Condition ii) guarantees that the fibers of $X \to S$ are all supersingular. Indeed, if X is a non-supersingular abelian variety of dimension g, then for any isogeny $X \to X'$, we have $a(X') < g$, hence the p-rank of $X'[p] \cap X'[F^2]$ is at most $2g - 1$. By induction we see the p-rank of

$$(13.3.1) \qquad \bigcap_{0 \leq i \leq g(g-1)/2+1} X[p^{g(g-1)/2+1-i}F^{2i}]$$

is at most

$$(13.3.2) \qquad \frac{(g^2 - g + 2)(2g - 1)}{2} < \frac{g(2g^2 - 3g + 5)}{2}.$$

13.4. Flat subgroup schemes after base change.

In the following we will give another definition of strictly supersingular abelian schemes which however turns out to be equivalent to Definition 13.3.

Lemma. *Let S be a noetherian scheme with only one associated point. Let $G \to S$ be a finite group scheme which is generically flat of rank r. Then there is a proper epimorphism $S' \to S$, which is an isomorphism over an open dense subscheme $U \subset S$, such that $G \times_S S'$ has a flat closed subgroup scheme of rank r. Furthermore, we may also require that S' has only one associated point.*

Proof. By Fact 2.2, the following functor

$$((S\text{-schemes})) \longrightarrow ((\text{sets}))$$
$$T \mapsto \{\text{flat closed subgroup schemes of } (G \times_S T) \text{ of rank } r \text{ over } T\}$$

is represented by a relatively projective S-scheme, say $f : S'' \to S$ together with $H \subset G \times_S S''$. Since $G \to S$ is generically flat, there is an open dense subscheme $U \subset S$ such that $G|_U$ is flat of rank r over U. Hence $f|_{f^{-1}(U)}$ is an isomorphism. Since f is proper and S has only one associated point, this shows f is an epimorphism. We can take a closed subscheme of $S' \subset S''$ such that S' has only one associated point and $f|_{S'}$ is also an epimorphism.   Q.E.D.

13.5. Abelian schemes which are constantly superspecial after an epimorphic base change.

Lemma. *Let S be a noetherian F_p-scheme with only one associated point. Let $E^g \times S \to X$ be an isogeny of abelian schemes over S ($g > 1$). If $X[p] = X[F^2]$, then there is a finite epimorphism $S' \to S$ such that $X \times_S S' \cong E^g \times S'$.*

Proof. When S is integral, see [45, Corollary 4.3].
    For general S, there is $n \geq 0$ such that F_S^n factors through S_{red}, inducing $f : S \to S_{\text{red}}$. Let $X_1 = X \times_S S_{\text{red}}$. Then there is a finite epimorphism $S_1 \to S_{\text{red}}$ such that $X_2 = X_1 \times_{S_{\text{red}}} S_1 \cong E^g \times S_1$. Let $S' = S \times_{S_{\text{red}}} S_1$. Then $S' \to S$ is an epimorphism. Let $X' = X \times_S S'$. Then we have

$$(13.5.1) \qquad X'^{(p^n)} \cong X^{(p^n)} \times_S S' \cong X_2^{(p^n)} \times_f S \cong E^g \times S'.$$

88

If $n > 0$, then by the assumption we have $X'^{(p^{n-1})}[p] = X'^{(p^{n-1})}[F^2]$, hence

$$(13.5.2) \qquad \ker(V : X'^{(p^n)} \to X'^{(p^{n-1})}) = X'^{(p^n)}[F] \cong E^g[F] \times S'$$

Therefore $X'^{(p^{n-1})} \cong (E^{(p)})^g \times S'$. Note that $E^{(p)} \cong E$. The lemma is then proved by inverse induction on n.   Q.E.D.

13.6. Strictly supersingular abelian schemes after epimorphic base changes.

Proposition. *Let S be a noetherian $\mathbf{F}_p$-scheme with only one associated point ξ. Let X be a strictly supersingular abelian scheme of relative dimension g over S. Then there is an epimorphism $f : S' \to S$, where S' has only one associated point ξ', such that*

i) *there is a "minimal isogeny" $\rho : E^g \times S' \to X \times_S S'$, in the sense that the generic fiber of ρ is minimal;*

ii) *for any choice of non-negative integers m, n, the group scheme*

$$(13.6.1) \qquad H_{mn} := (X[p^m] \bigcap_{0 \leq i \leq m-1} X[p^i F^{2m+n-2i}]) \times_S S'$$

has a flat subgroup scheme whose rank is equal to the rank of $(H_{mn})_{\xi'}$;

iii) *there is a filtration $E^g \times S' = X_{g-1} \to X_{g-2} \to ... \to X_0 \cong X \times_S S'$ whose factors are α-groups;*

iv) *there is an open dense subscheme $U \in S$ over which f is finite and*

$$(13.6.2) \qquad \ker(X_{g-1} \to X_n) \times_S U = (E^g[F^{g-1-n}] \times f^{-1}(U)) \cap (\ker(\rho) \times_S U)$$

for each n $(0 \leq n < g)$.

Proof. First, using the argument of Lemma 13.5 one shows there is $m \geq 0$ and an epimorphism $T_0 \to S$ such that there is an isogeny $E^g \times T \to X^{(p^m)} \times_S T_0$. By Lemma 13.4, for each $0 < n \leq g$ there is an epimorphism $T_n \to S$ such that $(\bigcap_{0 \leq i \leq n} X[p^{n-i} F^{2i}]) \times_S T_n$ has a flat subgroup scheme whose p-rank is equal to the p-rank r_n of $\bigcap_{0 \leq i \leq n} X_\xi[p^{n-i} F^{2i}]$, and

$$(13.6.3) \qquad (X[p^{g-1}] \cap \bigcap_{1 \leq i \leq g-1} X[p^{g-1-i} F^{2i+n}]) \times_S T_n$$

has a flat subgroup scheme whose p-rank is equal to the p-rank r'_n of

$$(13.6.4) \qquad X_\xi[p^{g-1}] \cap \bigcap_{1 \leq i \leq g-1} X_\xi[p^{g-1-i} F^{2i+n}].$$

Note that if $T \to S$ and $T' \to S$ are epimorphisms, so is $T \times_S T' \to S$. Hence we can choose an epimorphism $f : S' \to S$ such that

a) there is an isogeny $\rho' : E^g \times S' \to X \times_S S'$;

b) for any $n \leq g$, the group scheme $(\bigcap_{0 \leq i \leq n} X[p^{n-i}F^{2i}]) \times_S S'$ has a flat subgroup scheme H_n of p-rank r_n, and $(X[p^{g-1}] \cap \bigcap_{1 \leq i \leq g-1} X[p^{g-1-i}F^{2i+n}]) \times_S S'$ has a flat subgroup scheme H'_n of p-rank r'_n;

c) $S' \to S$ is generically finite, and S' has only one associated point ξ'.

Note that $r_g - r_{g-1} = 2g$. Let $Y_n = X \times_S S'/H_n$. Then the generic fiber of Y_{g-1} is superspecial. Note that there is an open dense subscheme $U \subset S$ such that $H_n \times_S U = (\bigcap_{0 \leq i \leq n} X[p^{n-i}F^{2i}]) \times_S U'$, where $U' = f^{-1}(U)$. Hence

$$(13.6.5) \qquad \ker(Y_{g-1} \to Y_g)|_{U'} = Y_{g-1}[F^2]|_{U'} = Y_{g-1}[p]|_{U'}.$$

Since S' has only one associated point, this shows $Y_{g-1}[F^2] = Y_{g-1}[p]$. Replacing S' via a generically finite epimorphism we will get $Y_{g-1} \cong E^g \times S'$ by Lemma 13.5.

Note that $p_{X \times_S S'}^{g-1}$ factors through Y_{g-1}, giving an isogeny $E^g \times S' \to X \times_S S'$, which is minimal since it is minimal at the generic fiber. This proves i).

Note that over U, the minimal isogeny ρ is equivalent to the canonical projection

$$(13.6.6)$$
$$E^g \times U' \cong X \times_S U'/(\bigcap_{0 \leq i \leq g-1} X[p^{g-1-i}F^{2i}] \times_S U') \to (X/X[p^{g-1}]) \times_S U' \cong X \times_S U'.$$

Hence for any $n > 0$, we have

$$(13.6.7) \qquad E^g[F^n] \times U' \cap \ker(\rho) \times_S U \cong (H'_n/H_{g-1}) \times_S U.$$

Let $X_n = X \times_S S'/H'_{g-1-n}$. Then $X_{g-1} \cong Y_{g-1} \cong E^g \times S'$ and $\ker(X_n \to X_{n-1}) \cong H'_{g-n}/H'_{g-1-n}$ which is an α-group. Thus iii) and iv) are proved.

For any $n \geq 0$, let

$$(13.6.8) \qquad G'_n := \ker(\rho) \cap X_{g-1}[F^n] = \begin{cases} \ker(X_{g-1} \to X_{g-1-n}) & n < g \\ \ker(X_{g-1} \to X_0) & n \geq g \end{cases}$$

and $G_n := X_{g-1}[F^n]/G'_n$ which is a flat subgroup scheme of X_0. Let $\phi_n : X_0 \to X_0/G_n$ be the projection.

We now prove ii). First we consider the case when $m \geq g - 1$. Let $G = (\bigcap_{0 \leq i \leq m} X[p^i F^{2m+n-2i}]) \times_S S'$. Then $H_{mn} = X_0[p^m] \cap G$. It is easy to check that $G_{2m+n} \subset G$ and $G_{\xi'} = (G_{2m+n})_{\xi'}$. Note that

$$(13.6.9) \qquad X_0/G_{2m+n} \cong X_{g-1}/X_{g-1}[F^{2m+n}] \cong E^g \times S'.$$

Note also that $X_{g-1}/X_{g-1}[p^m] \to X_0/X_0[p^m]$ is a minimal isogeny, and $\phi_{g-1-n} \circ p^m : X_{g-1} \to X_0/G_n$ factors through X_0/G_{2m+n}. This induces an epimorphism $\psi : G_{2m+n} \to G_n$. Hence H_{mn} has a flat subgroup scheme isomorphic to $\ker(\psi)$. It is easy to check that $\ker(\psi)_{\xi'} \cong (H_{mn})_{\xi'}$.

Suppose $m < g - 1$. Again we have $\ker(\psi) \subset H_{mn}$. To look for a flat subgroup scheme of H_{mn}, it is enough to look for a flat subgroup scheme of $H' := H_{mn}/\ker(\psi)$, which is a subgroup scheme of X_0/G_{2m+n}.

Note that F^m is a factor of p^m and each $p^i F^{2m+n-2i}$ $(0 \leq i < m)$, and $X_{g-1}^{(p^m)} \to X^{(p^m)}$ is a minimal isogeny. It is then easy to see that $\phi_{m+n}^{(p^m)} \circ F^m :$

90

$X_0 \to (X_0/G_{m+n})^{(p^m)}$ factors through X_0/G_{2m+n}. Let $G' = \ker(X_0/G_{2m+n} \to (X_0/G_{m+n})^{(p^m)})$. Then $G' \subset H'$. It is easy to check by a calculation on Dieudonné modules that $G'_{\xi'} \cong H'_{\xi'}$. Q.E.D.

13.7. A reason for the choices of intersections of $X[p^m F^n]$'s in Definition 13.3.

Remark. We have in fact proved that i), iii) and iv) together imply ii). Note that i), iii) and iv) hold when $X \to S$ is the end of a generically rigid FTQ.

However, we can not hope in general that any intersection of the $X[p^m F^n]$'s is generically flat when X is strictly supersingular. For example, consider the case of 9.4 ($g = 3$). We can take X, Y, Z which are linearly independent over F_{p^2} but are linearly dependent over F_{p^4}. Let $\xi \in \mathcal{P}_{3,\eta}$ be such a closed point with ideal sheaf $\mathcal{I}$ and $S \subset \mathcal{P}_{3,\eta}$ be defined by $\mathcal{I}^2$. Let X_S be the restriction of the end of the universal PFTQ over $\mathcal{P}_{3,\eta}$ on S. Then $X_S \to S$ is strictly supersingular and $a((X_S)_\xi) = 1$, but $X_S[p^2] \cap X_S[F^4]$ is not flat.

That's why we only have some special intersections of $X[p^m F^n]$'s in the definition of strict supersingularity.

13.8. Another definition of strictly supersingular abelian schemes.

Corollary. *Definition 13.3 is equivalent to the following:*

Definition. Suppose first that S has only one associated point ξ. An abelian scheme X over S is called *strictly supersingular* if

  i) the generic fiber X_ξ is supersingular;
  ii) for any choice of non-negative integers m, n, there is an epimorphism $f : S' \to S$, where S' has only one associated point ξ', such that the group scheme

$$(13.8.1) \qquad H_{mn} := (X[p^m] \cap \bigcap_{0 \le i \le m-1} X[p^i F^{2m+n-2i}]) \times_S S'$$

  has a flat subgroup scheme whose rank is equal to the rank of $(H_{mn})_{\xi'}$.

For general S, let $\mathcal{I}_1 \cap ... \cap \mathcal{I}_s$ be a (chosen) primary decomposition of $(0) \subset O_S$, and $S_i \subset S$ be the closed subscheme defined by $\mathcal{I}_i$ $(1 \le i \le s)$. Then we say $X \to S$ is *strictly supersingular* if each $X \times_S S_i \to S_i$ is so.

Proof. From Proposition 13.6.ii) we immediately see that Definition 13.3 implies the above definition (it is enough to check the case when S has only one associated point). On the other hand, it is obvious that the above definition implies Definition 13.3. Q.E.D.

13.9. FTQ structures after base change.

Corollary. *Under the condition of Proposition 13.6, we can choose S' and U such that there is an FTQ over S' ending at $X \times_S S'$ which is rigid over $U' = S' \times_S U$.*

Furthermore, if $X \to S$ has a principal polarization η_X, then there is a PFTQ (with respect to some η) over S' ending at $\{X \times_S S', \eta_X \times_S \mathrm{id}_{S'}\}$ which is rigid over U'.

Proof. For the first statement, it is enough to prove the following claim:

> Let S be a noetherian k-scheme with only one associated point ξ and $X \to S$ be an abelian scheme of dimension g. Suppose there is an isogeny $\rho : E^g \times S \to X$ such that
>
> i) $\ker(\rho) \cap E^g[F^n] \times S$ has a flat subgroup scheme H_n $(0 \le n \le g-1)$;
> ii) $(\ker(\rho) \cap E^g[F^n] \times S)_\xi \cong (H_n)_\xi$ $(0 \le n \le g-1)$;
> iii) if r_n is the α-rank of the α-group H_n/H_{n-1} $(1 \le n \le g-1)$, then $r_1 > 0$ and $r_{n-1} < r_n$ unless $r_{n-1} = g$.
>
> Then there is an endomorphism ϕ of $E^g \otimes k$ such that $\rho \circ (\phi \times \mathrm{id}_S) : E^g \times S \to X$ induces a rigid FTQ structure over some open dense subscheme $U \subset S$ ending at $X \times_S U$.

Proof of the claim. Take an elliptic curve $E \subset E^g$ such that $E \times S \to X$ is a closed immersion over some open dense subscheme $U' \subset S$. Let $X' = X \times_S U'/E \times U'$. Then there is an induced isogeny $\rho' : E^{g-1} \times U' \to X'$. If $r_1 = 1$, then ρ' factors through $F_{E^{g-1}} \times \mathrm{id}_{U'}$ and we replace ρ' by the induced isogeny $(E^{g-1}/E^{g-1}[F]) \times U' \to X'$. By induction on g, there is an endomorphism ϕ' of E^{g-1} such that $\rho' \circ (\phi' \times \mathrm{id}_S)$ induces a rigid FTQ structure over some open dense subscheme $U \subset U'$ ending at $X' \times_{U'} U$. Let ϕ be the pull-back of $\phi' \circ F$ via the projection $E^g \to E^{g-1}$. One then easily checks that $\rho \circ (\phi \times \mathrm{id}_S)$ induces a rigid FTQ structure over U ending at $X \times_S U$.

The proof of the last statement is similar using the trick of Proposition 6.3, and is left to the reader.   Q.E.D.

13.10. The lower bound of the p-rank.

Remark. The p-rank of $X[p^m] \bigcap_{0 \le i \le m-1} X[p^i F^{2m+n-2i}]$ is minimal if and only if $a(X) = 1$. In this case we denote the p-rank by r_{mn}.

13.11. Supersingular abelian schemes.

Definition. An abelian scheme $X \to S$ is called supersingular if there is a strictly supersingular abelian scheme $X' \to S'$ and a morphism $S \to S'$ such that $X \cong X' \times_{S'} S$.

Remark. Under this definition, any base change of a supersingular abelian scheme is obviously supersingular. It is also easy to see that a supersingular abelian scheme over a reduced base is strictly supersingular. However, a base change of a strictly supersingular abelian scheme may not be strictly supersingular. For example, take a non-constant family X of polarized supersingular abelian varieties over $S = \mathbb{A}^1_k = \mathrm{Spec}(k[t])$ whose fiber at 0 is superspecial. Let $S' = \mathrm{Spec}(k[t]/(t^2)) \subset S$ and $X' = X \times_S S'$. Then it is easy to see that $X' \to S'$ is not strictly supersingular.

13.12. Define the supersingular locus as a moduli.

We are going to give a *new* definition of supersingular locus.

Theorem. *For any $n > 2$ ($p \nmid n$), there is a closed subscheme $S'_{g,d,n} \subset A_{g,d,n} \otimes F_p$ representing the following functor*

$$\mathfrak{s} : ((F_p\text{-schemes})) \longrightarrow ((\text{sets}))$$
$$T \mapsto \{\text{supersingular abelian schemes of dimension } g \text{ over } T$$
$$\text{with a polarization of degree } d^2 \text{ and a level } n\text{-structure}\}.$$

Furthermore, the universal abelian scheme over $S'_{g,d,n}$ is strictly supersingular.

Proof. Let X be the universal abelian scheme over $A_{g,d,n}$ and

$$(13.12.1) \quad H_{mt} := (X[p^m] \bigcap_{0 \le i \le m-1} X[p^i F^{2m+t-2i}]) \quad (0 \le m, t \le \frac{g(g-1)}{2} + 1)$$

Let $[s]$ be a choice of integers $s_{mt} \ge r_{mt}$ ($0 \le m, t \le g(g-1)/2 + 1$). By flattening straightification, the functor

$$((F_p\text{-schemes})) \longrightarrow ((\text{sets}))$$
$$T \mapsto \{\text{supersingular abelian schemes of dimension } g \text{ over } T$$
$$\text{with a polarization of degree } d^2 \text{ and a level } n\text{-structure}$$
$$\text{such that } H_{mt} \times_{A_{g,d,n}} T \text{ is flat of } p\text{-rank } s_{mt} \; \forall m, t\}$$

is represented by a locally closed subscheme $U_{[s]} \subset A_{g,d,n} \otimes F_p$. Let $V_{[s]} \subset A_{g,d,n}$ be the scheme-theoretic closure of $U_{[s]}$, then the associated points of $V_{[s]}$ are all in $U_{[s]}$. Let $S'_{g,d,n}$ be the scheme-theoretic union of such $V_{[s]}$'s.

We now show that $S'_{g,d,n}$ represents $\mathfrak{s}$. Clearly $X \times_{A_{g,d,n}} S'_{g,d,n}$ is strictly supersingular. On the other hand, let $Y \to S$ be a strictly supersingular abelian scheme of dimension g with a polarization of degree d^2 and a level n-structure. For simplicity we may assume S has only one associated point ξ. Let $f : S \to A_{g,d,n}$ be the canonical morphism. Let s_{mt} be the p-rank of $(H_{mt})_{f(\xi)}$ ($0 \le m, t \le g(g-1)/2+1$). By the definition of $U_{[s]}$, there is a non-empty open subscheme $U \subset S$ such that $f|_U$ factors through $U_{[s]}$. Hence f factors through $V_{[s]}$.

Finally, by the construction of $V_{[s]}$ we see that the universal abelian scheme over $V_{[s]}$ is strictly supersingular, hence the universal abelian scheme over $S'_{g,d,n}$ is strictly supersingular by Definition 13.3.   Q.E.D.

Remark. Note that for any $n|m$ there is an étale cover $S'_{g,1,m} \to S'_{g,1,n}$. By this we can define coarse moduli schemes of supersingular abelian schemes $S'_{g,d,2} \subset A_{g,d,2}$ (for $p \ne 2$) and $S'_{g,d} = S'_{g,d,1} \subset A_{g,d}$.

13.13. A criterion of supersingularity.

Corollary. *An abelian scheme $X \to S$ of dimension g with a polarization of degree d^2 and a level n-structure ($p \nmid n$) is supersingular iff the canonical morphism $S \to A_{g,d,n}$ factors through $S'_{g,d,n}$.*

13.14. Define the supersingular locus in $\mathcal{A}_{g,1,n}$ as a moduli.

Theorem. *The moduli scheme $\mathcal{S}'_{g,1,n}$ $(n > 2, p \nmid n)$ defined in Theorem 13.12 (for $d = 1$) is (geometrically) reduced. Hence $\mathcal{S}'_{g,1,n} \cong \mathcal{S}_{g,1,n}$, the supersingular locus defined in 1.10.*

Proof. Since $\mathcal{S}'_{g,1,n}$ is an F_p-scheme, to show $\mathcal{S}'_{g,1,n}$ is geometrically reduced it is enough to show $\mathcal{S}'_{g,1,n}$ is reduced. By Theorem 13.12, it is enough to show that for any F_p-scheme S of finite type with only one associated point and any principally polarized strictly supersingular abelian scheme $X \to S$ of relative dimension g with a level n-structure, the induced morphism $S \to \mathcal{S}'_{g,1,n}$ factors through $(\mathcal{S}'_{g,1,n})_{\mathrm{red}}$.

By Corollary 13.9, there is an epimorphism $S' \to S$ such that $X \times_S S'$ is the end of a PFTQ, where S' has only one associated point. Hence there is an induced morphism $S' \to \mathcal{P}_{g,\eta}$ for some η. Since the PFTQ over S' is generically rigid, the image of S' in $\mathcal{P}_{g,\eta}$ intersects $\mathcal{P}'_{g,\eta}$.

Let $X_{g-1} \to ... \to X_0$ be the universal PFTQ over $\mathcal{P}_{g,\eta}$. Then we can take an étale cover $T \to \mathcal{P}_{g,\eta}$ such that $Y := X_0 \times_{\mathcal{P}_{g,\eta}} T$ has a level n-structure. Let $S'' = S' \times_{\mathcal{P}_{g,\eta}} T$. Let $T' \subset T$ be the inverse image of $\mathcal{P}'_{g,\eta}$ in T and $U \subset S''$ be the inverse image of T' under $S'' \to T$.

We choose a level n-structure of Y compatible with the level n-structure of $X \times_S S''$. This induces a morphism $T \to \mathcal{A}_{g,1,n}$. Note that T' is reduced, hence $T' \to \mathcal{A}_{g,1,n}$ factors through $(\mathcal{S}'_{g,1,n})_{\mathrm{red}}$. Since the induced morphism $U \to \mathcal{A}_{g,1,n}$ factors through T', it factors through $(\mathcal{S}'_{g,1,n})_{\mathrm{red}}$ also.

Note that $S'' \to S$ is an epimorphism and $S'' \to \mathcal{A}_{g,1,n}$ factors through S, and $U \to S$ is an epimorphism over an open dense subscheme $U' \subset S$. Hence the induced morphism $U' \to \mathcal{A}_{g,1,n}$ factors through $(\mathcal{S}'_{g,1,n})_{\mathrm{red}}$. Finally, since S has only one associated point, the induced morphism $S \to \mathcal{A}_{g,1,n}$ factors through $(\mathcal{S}'_{g,1,n})_{\mathrm{red}}$ also.   Q.E.D.

Remark. From Remark 13.12 we see that $\mathcal{S}'_{g,1,2}$ and $\mathcal{S}'_{g,1}$ are also reduced, hence $\mathcal{S}'_{g,1,n} \cong \mathcal{S}_{g,1,n}$ holds for $n \leq 2$ also. Thus we get:

Corollary. *For any g and n, the supersingular locus $\mathcal{S}_{g,1,n}$ can be defined as the coarse moduli scheme of principally polarized supersingular abelian schemes of relative dimension g with level n-structure, which is a fine moduli scheme when $n > 2$.*

13.15. A sufficient condition for strict supersingularity.

Corollary. *Let $X \to S$ be a principally polarized supersingular abelian scheme. If the fibers of X over the associated points of S are all supergeneral, then $X \to S$ is strictly supersingular.*

13.16. Another reason for the choices of intersections of $X[p^m F^n]$'s in Definition 13.3.

Example. If X is a supergeneral abelian variety of dimension 2, then $X[p^n] \cap X[F^{2n}]$ has p-rank $4n - 1$. Hence for a supersingular abelian scheme $X \to S$ (under any

94

possible definition), we can only hope that $X[p^n] \cap X[F^{2n}]$ has a flat subgroup scheme of p-rank $4n - 1$.

Let E be a supersingular elliptic curve over $K = \mathsf{F}_p$. Let $G = E[p]$. Then the group scheme structure of G is given by

$$(13.2.1) \quad G \cong \mathrm{Spec}K[x]/(x^{p^2}), \quad m^*(x) = x \odot 1 + 1 \otimes x + \sum_{i=1}^{p-1} (-1)^i i x^{ip} \otimes x^{(p-i)p}.$$

Let $S = \mathrm{Spec}K[\epsilon]/(\epsilon^2)$. Let $H \subset G \times S$ be defined by the ideal $(x^p - \epsilon x)$. Note that H is flat of rank p over S. Let $Y = E \times S$, $Y' = Y/H$ and $X = Y \times_S Y'$. It is easy to see that for any $n > 0$, $H_n = X[p^n] \cap X[F^{2n}]$ has a flat subgroup scheme of p-rank $4n - 1$. However H_n is not generically flat, because the p-rank of the generic fiber of H_n is $4n$, but H_n does not have a flat subgroup scheme of p-rank $4n$.

Clearly $X \to S$ should not be included in any possible definition of supersingular abelian schemes (because $Y'[p] \neq Y'[F^2]$, hence is not supersingular). This example gives a reason for the requirement of generic flatness in Definition 13.3.i).

Remark. The above example also gives one of the reasons of the complexity in Definition 13.3: It is not enough to put conditions on $X[p^n] \cap X[F^{2n}]$ only, as in the case $g = 1$ (see Remark 13.7 also).

Similarly, to give a definition it is not enough to put conditions of the form $p^r|V^s$ only (though one can use conditions of this form to define supersingular abelian schemes over a *reduced* base, see [38, Corollary 2.3.2] and [59, Theorem 1.2]). For example, take *any* abelian scheme $X \to S = \mathrm{Spec}(k[t]/(t^2))$ whose closed fiber is superspecial. Then $X^{(p)}$ is constant and superspecial, hence for any $r > 0$ there is an isomorphism f such that $V^{2r}_{X^{(p)}/S} = p^r f$, thus $p^r|V_{X/S} \circ V^{2r}_{X^{(p)}/S} = V^{2r+1}_{X/S}$.

A. Appendix: Some historical remarks

The main result of this book is the study of families of supersingular abelian varieties, in particular the dimension of their moduli spaces, and the number of components of the supersingular locus in the moduli space of polarized abelian varieties. This study was started in the fundamental paper by Deuring ([9]), where supersingular elliptic curves were considered. In the Dieudonné-Manin classification as presented in [48], various types of abelian varieties were studied. By the work of Tate, Eichler, Deligne, Shioda and Oort we have various characterizations of supersingular abelian varieties. In a paper by Oda and Oort ([61]) we find a conjecture what should be the dimension of the locus of principally polarized supersingular abelian varieties. For $g \leq 3$ this was proved by Katsura and Oort ([31], [35], [36]). In [45] the study was carried on by Li. Finally, in this book we prove the conjecture stated in [61], and we complete the study of the dimension and the number of components of the supersingular locus for polarized abelian varieties of arbitrary dimension.

In this Appendix we give a (probably incomplete) review of some recent results on supersingular elliptic curves and supersingular abelian varieties.

A.1. Singular j-values of elliptic curves.

Classically, the j-invariant of an elliptic curve is called *singular* in case that elliptic curve (over $\mathbb{C}$) has more endomorphisms than only multiplication by an integer (cf. [102, Section 115]). It is already known for a long time that singular j-values are algebraic integers (cf. [102, 115.VI]). Some of these values were computed, and Weber remarks "Bei diesen Zahlen ist die Zerlegbarkeit in verhältnismässige kleine Primzahlen bemerkenswert."(cf. [102, Section 125, p.462]) which was explained and proved again in [22]. For a table giving the 13 rational singular j-values see [88, p.192]. In [22] we also see how to decide which prime factors divide a singular j-value, while this was adapted for the case of singular Drinfeld modules in [10]. For the role played by singular j-values for the theory of abelian extensions of imaginary quadratic fields see [93, Chapter 6].

A.2. Supersingular elliptic curves.

H. Hasse discovered that it can happen for an elliptic curve E over a field of prime characteristic that its endomorphism ring $\mathrm{End}(E)$ is of rank 4 (and in that case the algebra is non-commutative). In Deuring [9] we find the proof that indeed this occurs for any prime characteristic. In [9, p.249] we find the terminology of "supersingular" j-value for the case that the endomorphism ring $\mathrm{End}(E \otimes k)$ is of rank 4, where k is an algebraically closed field, e.g. $k = \bar{\mathbb{F}}_p$, the algebraic closure of the prime field. It has become the custom to call an elliptic curve with this property a *supersingular elliptic curve* (although the curve is not singular at all). We have the

96

following possibilities for an elliptic curve E defined over a field K of characteristic p :

- The value $j(E)$ is transcendental over $\mathbf{F}_p$, this is the case if and only if $\text{End}(E) = \mathbf{Z}$.

- For $K \subset \bar{K} \subset k$ we have that $R = \text{End}(E \otimes k)$ is of rank 2 over $\mathbf{Z}$. In this case the j-value is called *singular*, $D = \text{End}^0(E \otimes k) = R \otimes \mathbf{Q}$ is an imaginary quadratic field in which p splits, all such fields can be obtained in this way, and $j(E)$ is algebraic over the prime field.

- There is a field extension $K \subset L$ such that $R = \text{End}(E \otimes L)$ is non-commutative. In this case we say E is *supersingular*. Note that in this case $D = R \otimes \mathbf{Q}$ is the quaternion algebra over $\mathbf{Q}$ ramified at ∞ (i.e. D is a definite quaternion algebra) and at p (and unramified at all other places of $\mathbf{Q}$), and R is a maximal order in D. Deuring denotes this algebra by $D = Q_{\infty,p}$. In case E is supersingular, we have $j(E) \in \mathbf{F}_{p^2}$, and any two supersingular elliptic curves over an algebraically closed field k are isogenous. The number of supersingular j-invariants equals the class number $h = h_p$ of $Q_{\infty,p}$, which was computed by Eichler, see Example 9.1. Later Igusa gave a beautiful geometric proof of this computation, see [29].

A.3. Abelian varieties with sufficiently many complex multiplications.

As we have seen: an elliptic curve over a field of positive characteristic has "complex multiplications" if and only if it can be defined over a finite field. This has been generalized as follows:

Tate has shown that an abelian variety defined over a finite field has "sufficiently many complex multiplications" (some people say: "is of CM-type", see 0.6 for the definition, also see [97], [98]).

The converse is almost but not quite true, in fact Grothendieck proved the following theorem: Suppose an abelian variety X over a field K has sufficiently many complex multiplications, then

- if char(K) = 0, then X can be defined over a number field (see [94, 12.4, Proposition 26], note that the proof of this fact by Grothendieck in [66] is different form the one in [94]),

- in general, there is an isogeny $X \sim Y$ defined over some finite extension of K such that Y can be defined over a finite extension of the prime field (cf. [66], and note that if char(K) = $p > 0$, there exists such examples in all dimensions $g > 1$, where X cannot be defined over a finite field).

Note that any two supersingular elliptic curves defined over the same algebraically closed field are isogenous. This was proved by Deuring in [9] via lifting techniques. In [98] Tate shows that abelian varieties over a finite field with isogenous ℓ-divisible groups (or with equal Weil numbers) are isogenous, and this gives a new proof of the above fact. In [98] there is an ℓ-adic description of homomorphisms of abelian varieties, while in [101] we find the proof by Tate of the p-adic version of this theorem.

A.4. Classification of abelian varieties over finite fields.

One can try to classify abelian varieties over finite fields. Up to isogeny this is completely done by Honda-Serre-Tate via the method of Weil numbers (cf. [98]). Classification up to isomorphism is more difficult. For elliptic curves Waterhouse gives a complete answer, in particular [100, Theorem 4.1.2] describes all supersingular elliptic curves with all endomorphisms defined over $\mathbf{F}_q$, and [100, Theorems 4.1.3-5] gives all supersingular elliptic curves with as endomorphism algebra an imaginary quadratic field. For some further information concerning (ordinary) abelian varieties over finite fields, see [5], [82], [26].

Just to mention two questions:

- Given an abelian variety over a finite field K, is it K-isogenous to an abelian variety which admits a principal polarization?

- Given $g \in \mathbf{Z}_{>0}$, does there exist an effective bound $N = N(g, q)$ such that any abelian variety of dimension g defined over $\mathbf{F}_q$ admits a polarization of degree at most N?

For the first one, Howe has some results (see [28]); for the second one we do not know the answer.

A.5. Newton polygons.

In the fundamental paper [48], Manin applied Dieudonné module theory, and we see that an abelian variety in positive characteristic can be distinguished by several finer invariants, such as p-rank, and Newton polygon (also called "formal isogeny type", see 0.6 and 1.4, cf. [6, pp. 86/87], [38, pp. 119-122], [75]). An abelian variety X in positive characteristic is called *supersingular* if its Newton polygon has all slopes equal to $1/2$. This is equivalent to saying that its p-divisible group $\varphi_p(X)$, or its formal group $\hat{X}$, is isogenous to $\hat{E}^g$, where E is a supersingular elliptic curve (see 1.4). We see the following curious phenomenon: in general a splitting of the formal group (or a splitting of the Lie algebra, or something like that) does not imply that the abelian variety in question is (isogenous to) a product, however:

- If X is supersingular, then there exists an isogeny $X \otimes k \sim E^g$, where E is a supersingular elliptic curve, and k is an algebraically closed field (cf. [67, Theorem 4.2]), but:

- For any formal isogeny type unequal to the supersingular one (and for every positive characteristic p) there exists an absolutely simple abelian variety having that formal isogeny type (cf. [44]).

Note that the endomorphism algebra of $X \otimes k$ has rank $(2g)^2$ over $\mathbf{Q}$ if and only if X is supersingular (cf. [97, Theorem 2.d]).

Remark that $a(X) = \dim(X)$ if and only if X is *superspecial*, i.e. isomorphic (over an algebraically closed field) to a product of supersingular elliptic curves (cf. [69]).

A.6. Polarized flag type quotients.

Supersingular abelian varieties are all isogenous with one fixed abelian variety. This gives the possibility to describe moduli via such isogeny correspondences. In Oda-Oort [61] this is worked out, and the notion of a "flag type quotient" was formulated (see Definition 3.2). It was proved that a *supergeneral* abelian variety (i.e. a supersingular one with $a(X) = 1$) is the end of a *unique* FTQ (see [61, Theorem 2.2]). It was realized the notion of a polarized flag type quotient (especially in case $g > 3$) should be quite subtle. However one could speculate on the number of conditions a polarization would impose, and in [61, pp. 615-616] we find the conjecture that the dimension of the moduli space of principally polarized abelian varieties should be equal to $[\frac{g^2}{4}]$ (which turns out to be correct, see Theorem 4.9).

In [36, Definition 4.2] we find an attempt to coin the notion of a polarized flag type quotient, and [36, Proposition 5.4] shows that every principally polarized supersingular abelian variety of dimension 3 is the end of a PFTQ (note that in some cases the flag type quotient is not unique). In dimension $g = 3$ already we see that the morphism Ψ mapping the moduli point of a PFTQ to the moduli point of its end contracts positive dimensional subvarieties (see (9.4.16)).

In Definition 3.6 we have refined this notion, and in particular we have introduced the notion of a *rigid* PFTQ. Most of the above paper hinges on the fact that every principally polarized supersingular abelian variety is the end of a rigid PFTQ, see Proposition 4.1 (and on the set of rigid PFTQ's the morphism Ψ is finite-to-one).

One could speculate on definitions for FTQs and PFTQs for abelian varieties having a Newton polygon which is not supersingular. Such a notion can be phrased, and it produces subsets of a Newton polygon stratum which together look like a fibration. For the supersingular stratum each component is one "fibre" (every supersingular abelian variety is isogenous to a fixed one), whereas for a non-supersingular stratum there is a continuous family of "fibres" (the isogeny classes of such polarized abelian varieties "have moduli"). We expect to come back to this interesting phenomenon.

A.7. Superspecial abelian varieties.

For an abelian variety of dimension $g \geq 2$ over an algebraically closed field of characteristic p :

$$a(X) = g \quad \Longleftrightarrow \quad X \sim E^g,$$

where E is a supersingular elliptic curve (one uses results of [67], [69], and results by Ogus [62] and by Deligne, see [95], see 1.6 for more details).

A.8. More refined definitions for supersingularity.

One can try to give a more refined definition for supersingularity, which could be used for defining the infinitesimal structure for that moduli space. In [49], in [6, IV.5, pp. 86-88], and in [38] we find such a characterization. The results [38, Theorem 2.3.1 and Corollary 2.3.2] by Grothendieck and Katz show that the supersingular locus (or another locus defined by giving a Newton polygon) is closed. However the "correct" scheme structure on this "moduli space" is not clear. Further attempts we find in [59], and [3], but also these results seem not to provide a canonical way of defining the desired scheme structure (see Remark 13.15).

In [12] we find several results on polarized supersingular abelian varieties, mass formulas, and especially results on "superspecial" curves, see [12, Theorem 1.1, p.165]. We mention the problem on [12, p.173]: "Does there exist a curve of genus 4 or 5 whose Jacobian is the product of supersingular elliptic curves for any $p \neq 2, 3$?"

A.9. Families of supersingular abelian varieties and the supersingular locus.

There exist in positive characteristic p a quite natural non-trivial family of super-singular abelian surfaces over $\mathbf{P}^1$: for $t = (a : b) \in \mathbf{P}^1$ consider the abelian surface $X_t := (E \times E)/t(\alpha_p)$ (e.g. cf. [67, 4.7 and 2.6], and see [69, p.35]). Note that for general t we have $a(X_t) = 1$. Moret-Bailly described how one can find a principal polarization on such an abelian scheme (cf. [51]), and therefore such families are called Moret-Bailly families. Moreover in [51, p.133, Proposition 2.5], we find the fact that the polar divisor $\mathcal{C} \to \mathbf{P}^1$ of such a family, which is generically a non-singular genus two curve has exactly $5p - 5$ singular fibres (and such a singular fibre is a join of two supersingular elliptic tails).

From this result one concludes immediately that for a prime number $p > 1500$ the supersingular locus $\mathcal{S}_{2,1} \subset \mathcal{A}_{2,1}$ is reducible: the number $h = h_p$ of supersingular j-invariants is roughly $p/12$, hence the number $h(h-1)/2$ of isomorphism classes of unordered pairs of supersingular elliptic curves is roughly $p^2/(2 \cdot 12 \cdot 12)$: in case this is larger than $5p - 5$ we see that $\mathcal{S}_{2,1}$ is the image of more than one Moret-Bailly family.

Of course this can be made much more precise, and in fact by purely geometric methods (using Moret-Bailly families, and the ramification behavior of the maps of these into the moduli space) one can show that $\mathcal{S}_{2,1}$ is reducible if and only if $p > 11$ (see [31], [35], especially see [35, Theorem 5.8], and see the examples worked out in [35, Section 8]).

The number of components of $\mathcal{S}_{2,1}$ can be determined for every p: one applies methods about quaternion hermitian forms as developed by Shimura (see [92]), and one shows that the number of geometric components of $\mathcal{S}_{2,1}$ equals $H_2(1,p)$ (see [35, Theorem 5.7]). Using the computation of Hashimoto and Ibukiyama ([25, II]), this yields a complete description of this number of geometrically irreducible components of $\mathcal{S}_{2,1}$ for every prime number p. Then one can study each component separately.

As explained in [36] (see (9.2.4.) above), for $g = 2$, and for each polarization η on E^2 with $\ker(\eta) = E^2[F]$, there is a group G_η which acts on $\mathbf{P}^1 \cong \mathcal{P}_{2,\eta}$, such that $\mathcal{P}_{2,\eta}/G_\eta$ is the normalization of a component of $\mathcal{S}_{2,1}$. In this way the approach just mentioned is refined. Ibukiyama showed in [32, Theorem 7.1] that these groups do occur, and we find a way to decide for every characteristic p which groups do appear.

It would be nice to give an explicit geometric description of the maps $\Psi_\eta : \mathcal{P}_{g,\eta} \to \mathcal{S}_{g,1}$ for every g and every p (and see Proposition 9.12 for a first approach to this question).

Please note the parallel between on the one hand for $g = 1$ the computation by Deuring of the number h_p of supersingular j-invariants in characteristic p using a class number computation by Eichler, and on the other hand the methods just described for $g = 2$. Reflecting on these developments and using [61, Theorem 2.2], it is clear that the number of geometric components of $\mathcal{S}_{g,1}$ which generically have $a = 1$ can be expressed as a certain class number, see Corollary 4.8. The main point of the present book is to show that all components of $\mathcal{S}_{g,1}$ have $a = 1$ generically!

Note that this is true for principally polarized supersingular abelian varieties (see Theorem 4.9.iii), also see [39, Theorem 7.2] and [58, Theorem 4.1]). Note that for non-principally polarized analogous questions are difficult and may depend on the stratum in consideration, see [58, Theorem 4.1] on the one hand, and [36, Remark 6.10] and Proposition 10.4.iii) on the other hand.

This is a particular case of the following fact: A principally polarized abelian variety (X_0, λ_0) in characteristic p can be deformed to (X, λ) such that the Newton polygons of X_0 and X are equal, and $a(X) \leq 1$ (cf. [75, p.388]). Note that for non-principally polarized abelian varieties such deformations in general do not exist.

One can ask whether the beautiful geometric proof by Igusa ([29]) has an analogue for higher g. In fact, for higher g there are two interesting aspects.

One is the geometric method just described, and present in Proposition 4.2: the supersingular points in $\mathcal{A}_{1,1}$ for $g = 1$ are replaced by the possible choices for η. Note however:

- for $g = 1$ the number of supersingular elliptic curves equals $h = h_p$ and the number of principal polarizations of each curve is one,

- but for $g > 1$, by the theorem mentioned in 1.6, for each prime number p there is only one superspecial abelian variety X of dimension g up to isomorphism, but the number of choices for η depends on g and on p,

where $g = \dim(X)$ and η is a polarization of X such that $\ker(\eta) = X[F^{g-1}]$. Furthermore, the number of such polarizations η up to equivalence is given by the class number $H_g(p, 1)$ if g is odd, and $H_g(1, p)$ in case g is even.

Another is the method developed in [31, pp. 132-138]: there one considers special one-dimensional families of abelian surfaces, which in a certain sense are very much like the Legendre family for elliptic curves. The supersingular points of such a family are, as in the case studied by Igusa, given by the zeros of a certain polynomial and also in this case the crucial information that these zeros all have multiplicity one comes from the fact that this function satisfies a hypergeometric differential equation, see [31, Proposition 1.14]. It might be interesting to see whether for higher g analogous results can be derived.

A.10. Further techniques on the classification of supersingular abelian varieties.

In 1985 the work of Li on the classification of supersingular abelian varieties provided some techniques, most of which have been used in this book (see [45, Sections 1-4]), including:

- A lemma which characterizes supersingular abelian varieties with a-number 1 (see Fact 5.6 and Fact 5.8);

- A study of some integer invariants of supersingular Dieudonné modules, which give geometric invariants of irreducible components of $S_{g,d}$ when $d > 1$ (see 12.2);

- A constructive way to recover an FTQ over a reduced base S from its Dieudonné crystal, which shows that the category of FTQs and the category of FTQs of Dieudonné crystals are almost equivalent (see Lemma 7.4 and Remark 7.5).

Li's paper also includes a deep theorem of Ogus (see [45, Lemma 4.1]), which says that for a supersingular abelian scheme X (i.e. an abelian scheme whose fibers are supersingular) over a *normal* base S, there exists an étale cover $\epsilon : S' \to S$ such that there is an isogeny $E^g \times S' \to X \times_S S'$ (where E is a supersingular elliptic curve). In particular, if X is a *superspecial* abelian scheme over a normal base S, then there is an étale cover $\epsilon : S' \to S$ such that $X \times_S S' \cong E^g \times S'$. On the other hand, there are examples to show the necessity of ϵ, i.e. there are superspecial abelian schemes of dimension $g > 1$ over some normal base S which are not isomorphic to $E^g \times S$ (see [69, Remark 10] and [45, Section 4, Example 1]).

A.11. The relation with the Torelli locus.

Suppose we want to study polarized supersingular abelian varieties satisfying an extra property, e.g. that it is a Jacobian. This amounts to studying the intersection of subsets in the moduli space (in this case the supersingular locus with the Torelli locus). Our experience is that such a question can be quite hard.

In [73] we find: for a fixed prime number, every component of the moduli space of hyperelliptic supersingular curves of genus $g = 3$ has dimension one; this amounts to intersecting in a 6-dimensional moduli space the 2-dimensional supersingular locus with the 5-dimensional Torelli locus. Already this innocent looking question turned out to be non-trivial, and the proof in [73] is quite indirect. It would be nice to have better methods for such questions.

Let $T_g^0 = j(\mathcal{M}_g) \hookrightarrow \mathcal{A}_{g,1}$, the open Torelli locus in genus g, and T_g be the Zariski closure of T_g^0 in $\mathcal{A}_{g,1}$, i.e. the closed Torelli locus. What is the intersection

$$T_g \bigcap S_{g,1} = ?$$

See [76, (4.7)] for a further discussion. In any case, we have no idea, no conjecture to offer what the dimension(s) of the components of the intersection above could be! Are the dimensions of the components of this intersection independent of p? Note that for g large, the sum $(3g - 3) + [g^2/4]$ of the dimensions of T_g and $S_{g,1}$ is less than $g(g + 1)/2 = \dim(\mathcal{A}_g)$; this could suggest that the intersection $T_g^0 \cap S_{g,1}$ is empty in such cases, however there are values of p and g (large) for which there exist irreducible supersingular curves:

- Using the methods of coding theory, van der Geer and van der Vlugt showed that for any g, there exists a non-singular supersingular curve of genus g in characteristic 2 (see [21, p.53]).

- Let p be an odd prime number. For any $n > 0$ there exists a non-singular supersingular curve of genus $g = (p^n - 1)/2$ in characteristic p (see [39, Corollary on p.208]); More generally, if the only digits appearing in the p-adic expansion of g are 0 and $(p - 1)/2$, then there exists a non-singular supersingular curve of genus g in characteristic p (see [21, p.53], this was also proved using the methods of coding theory).

In [12, Section 3], Ekedahl describes families of supersingular curves of genus 2.

A.12. Supersingular reductions.

Suppose given an elliptic curve E over a number field K. What can be said about the set Σ of places of K where E has good supersingular reduction? In [87, pp. 250-251] we see that Σ has a density, and this equals 0 iff and only if $\text{End}(E) = \text{End}(E \otimes \mathbf{C})$, and this density equals $1/2$ in case E has no CM over K, but acquires CM over an extension of K. More precise statements are known, e.g. cf. [41]. Elkies has shown that for every elliptic curve over a real number field the set Σ of supersingular reductions is infinite (cf. [14], [15]).

Analogous statements for higher dimensions for non-CM abelian varieties seem difficult. Ogus proved that for an abelian surface over a number field the set of ordinary primes is infinite. See [42] and [57] for some further information. However the following question seems open for $g > 2$:

– Let X be an abelian variety over a number field. Does there exist a place of good, ordinary reduction? (One expects that the set of such places has positive density.)

For abelian varieties with sufficiently many complex multiplications (abelian varieties "of CM-type") the situation is easier: The set of supersingular primes for a CM-type abelian variety over a number field is infinite (cf. [99, Corollary 3.1.3]).

However we have no idea what can be said in general. Just to mention some questions:

– Suppose E_1 and E_2 are two elliptic curves, both without CM, with different j-invariants. What can be said about the set of primes where both E_1 and E_2 are supersingular?

– Does there exist an abelian surface over a number field with no supersingular reduction?

– More generally: Fix an abelian variety X over a number field K, fix a Newton polygon α, let Σ_α be the set of finite places of K where X has good reduction such that the reduction has Newton polygon equal to α; what can be said about this set? Clearly one can give some examples where this set is empty. But general statements, conjectures, or expectations are not obvious to us.

For abelian varieties over a number field with endomorphism ring larger than $\mathbf{Z}$, one can decide that certain Newton polygons do not appear in the reduction modulo a prime. Here is a different example. R. Noot studied the one-dimensional Shimura variety of Hodge type for abelian varieties of dimension 4 as given in [106,

Section 4]. Most of these abelian varieties have endomorphism ring equal to $\mathbb{Z}$ (the Hodge variety is not of PEL-type, i.e. these abelian varieties are not characterized by their endomorphism ring). Noot showed (unpublished) that a Newton polygon not appearing in the following list does not appear in the reduction modulo a prime of a fiber in one of these Mumford families:

$(0,0) - (4,0) - (8,4)$ (the ordinary type), $(0,0) - (3,1) - (6,3) - (8,4)$.
$(0,0) - (4,1) - (8,4)$, $(0,0) - (8,4)$ (the supersingular type).

A.13. The structure of the other strata.

The supersingular locus $\mathcal{S}_{g,1}$ in the moduli space of *principally polarized* abelian varieties of dimension g in characteristic p is a special case of the more general situation where one considers a Newton polygon α (belonging to g), and where one studies the locus W_α. A stratum given by a p-rank is a special case of such a stratum. Questions concerning such subsets of the moduli spaces were studied in [58], in [75] and will be taken up in [13] and in [77]. We mention some of the results derived:

- For $g \geq 2$ *every stratum W_α is connected.* This follows from the main result of [13]. T. Ibukiyama had a proof for this in case of the supersingular locus in $g = 2$ (unpublished). This statement generalizes (and gives a new proof) of the theorem by Faltings and by Chai that the moduli space $\mathcal{A}_{g,1} \otimes \mathbb{F}_p$ is irreducible.

- *Every locus W_α which is not the ordinary one generically has $a = 1$, and its dimension can be computed directly from the Newton polygon* (see [75, Theorem 2.6]). This generalizes the fact that $\dim(\mathcal{S}_{g,1}) = [g^2/4]$ (at the same time it gives an "explanation" of this number), and it generalizes [39, Theorem 7] and [58, Theorem 4.1].

- Note that analogous statements are in many cases not correct if one considers non-principally polarized abelian varieties. E.g. see [36, Remark 6.10], where for the case $g = 3$ it is shown that there exists a component of the supersingular locus $\mathcal{S}_g$ of dimension 3 (which is $\neq [3^2/4]$), and there exists a component where the a-number generically is not equal to 1. This is studied in further detail in Chapter 10 and Chapter 12.

- We have seen above that the number of components of the supersingular locus (fix a degree of the polarization) can be given by a class number. For p large usually such a class number is big, and one concludes that in that case the relevant supersingular locus is reducible. However we have several good reasons to hope that the following is correct:

  – Let α be a Newton polygon which is not supersingular; then is the locus $W_\alpha \subset \mathcal{A}_{g,1} \otimes \mathbb{F}_p$ irreducible (?). This would finish the task of describing the number of components of all Newton polygon strata.

Two questions on which we have no idea what to expect:

– Given g and α, what is $\mathcal{T}_g \bigcap W_\alpha$? (It is clear that the generic point of $\mathcal{T}_g$ corresponds to an ordinary Jacobian.) Note that for g large the length of the graph of Newton polygons is longer than $3g - 3$, thus it is not true that these intersections are all non-empty ánd nested as given by the ordering of Newton polygons (cf. [39, corollary on p.214]). What can be said about the number of components of this intersections, of the various dimensions; for a given g does every α appear as a Newton polygon of a curve; are these answers independent of p? (See A.11 above.)

Does every Newton polygon appear as the Newton polygon of a Jacobian? We have no reasonable guess. We do not know whether the answer depends on p.

References

[1] P. Berthelot, L. Breen & W.Messing - *Théorie de Dieudonné cristalline II.* LNM 930, Springer-Verlag (1982).

[2] P. Berthelot & A. Ogus - *Notes on Crystalline Cohomology.* Princeton Univ. Press & Univ. Tokyo Press., Princeton (1978).

[3] C.-L. Chai - Every ordinary symplectic isogeny class in positive characteristic is dense in the moduli. Invent. Math. 121 (1995), 439-479.

[4] P. Deligne - Hodge cycles on abelian varieties. In: *Hodge Cycles, Motives and Shimura Varieties* (P. Deligne, J. Milne, A. Ogus, K-Y. Shih), LNM 900, Springer-Verlag (1982).

[5] P. Deligne - Variétés abéliennes ordinaires sur un corps fini. Invent. Math. 8 (1969). 238-243.

[6] M. Demazure - *Lectures on p-divisible Groups.* LNM 302, Springer-Verlag (1972).

[7] M. Demazure & P. Gabriel - *Groupes algébriques, I.* Masson, Paris and North-Holland Pub. Co., Amsterdam (1970).

[8] M. Demazure & A. Grothendieck - *Schemas en Groupes I-III (SGA 3).* LNM 151-153, Springer-Verlag (1970).

[9] M. Deuring - Die Typen der Multipikatorenringe elliptischer Funktionenkörper. Abh. Math. Sem. Hamburg 14 (1941), 197-272.

[10] D. R. Dorman - On singular moduli for rank 2 Drinfeld modules. Invent. Math. 110 (1992), 419-439.

[11] M.Eichler - Über die Idealkalssenzahl total definiter Quaternionenalgebren. Math. Z. 43 (1938), 102-109.

[12] T. Ekedahl - On supersingular curves and abelian varieties. Math. Scand. 60 (1987), 151-178.

[13] T. Ekedahl & F. Oort - Connected subsets of a moduli space of abelian varieties. (To appear).

[14] N. Elkies - The existence of infinitely many supersingular primes for every elliptic curve over Q. Invent. Math. 89 (1987), 561-567.

[15] N. Elkies - Supersingular primes for elliptic curves over real number fields. Compos. Math. 72 (1989), 165-172.

[16] G. van der Geer & M. van der Vlugt - Reed-Muller codes and supersingular curves.I. Compos. Math. 84 (1992), 333-367.

[17] G. van der Geer & M. van der Vlugt - Kloosterman sums and the p-torsion of certain Jacobians, Math. Ann. 290 (1991), 549-563.

[18] G. van der Geer & M. van der Vlugt - Supersingular curves of genus 2 over finite fields of characteristic 2. Math. Nachr. 159 (1992), 73-81.

[19] G. van der Geer & M. van der Vlugt - Artin-Schreier curves and codes. J. Algebra 139 (1991), 256-272.

[20] G. van der Geer & M. van der Vlugt - Curves over finite fields of characteristic 2 with many rational points. (To appear in C. R. Acad. Sci. Paris, 1993).

[21] G. van der Geer & M. van der Vlugt - On the existence of supersingular curves of given genus. J. reine angew. Math. 458 (1995), 53-61.

[22] B. H. Gross & D. Zagier - On singular moduli. Crelle, 35 (1985) 191-220.

[23] R. Hartshorne - *Algebraic Geometry*, GTM 52, Springer-Verlag (1977).

[24] K. Hashimoto - Class numbers of positive definite ternary quaternion hermitian forms. Proc. Japan Acad. 59 (1983), 490-493.

[25] K. Hashimoto and T. Ibukiyama - On the class numbers of positive definite binary quaternion hermitian forms, I. J. Fac. Sci. Univ. Tokyo, Sect IA, 27 (1980), 549-601; Part II, ibid. 28 (1981), 695-699; Part III, ibid. 30 (1983), 393-401.

[26] E. W. Howe - Kernels of polarizations of ordinary abelian varieties over finite fields. Ph.D. thesis, Berkeley (1993).

[27] E. W. Howe - Principally polarized ordinary abelian varieties over finite fields. Transact. AMS 347 (1995), 2361-2401.

[28] E. W. Howe - Bounds on polarizations of abelian varieties over finite fields. (To appear in J. angew. Math.).

[29] J-I. Igusa - Class number of a definite quaternion with prime discriminant. Proc. Nat. Acad. Sci. USA, 44 (1985), 312-314.

[30] S. Iyanaga (editor) - *The Theory of Numbers*. North-Holl. Math. Libr. 8, North-Holland Pub. Co. (1975).

[31] T. Ibukiyama, T. Katsura & F. Oort - Supersingular curves of genus two and class numbers. Compos. Math. 57 (1986), 127-152.

[32] T. Ibukiyama - On automorphism groups of positive definite binary quaternion hermitian lattices and new mass formula, Advanced Studies in Pure Mathematics 15, Kinokuniya Company and Academic Press (1989), 301-349.

[33] A.J. de Jong - Moduli of abelian varieties and Dieudonné modules of finite group schemes, Ph.D. thesis, Nijmegen (1992).

[34] A.J. de Jong - The moduli spaces of polarized abelian varieties. Math. Ann. 295 (1993), 485-503.

[35] T. Katsura & F. Oort - Families of supersingular abelian surfaces. Compos. Math. 62 (1987), 107-167.

[36] T. Katsura & F. Oort - Supersingular abelian varieties of dimension

two or three and class numbers. Adv. St. Pure Math. 10 (1987) (Algebr. Geom., Sendai, 1985; Ed. T. Oda), Kinokuniya Co., Tokyo Japan, and North-Holland Co., Amsterdam (1987).

[37] N. M. Katz - Serre-Tate local moduli. In: *Surfaces algébriques* (Sém. de géom. alg. d'Orsay 1976-78). LNM 868, Springer-Verlag (1981), Exp. V-bis, 138-202.

[38] N. Katz - Slope filtration of F-crystals. Journées Géom. Algébr. Rennes 1978, Vol. I. Astérisque 63, Soc. Math. France (1979), 113-164.

[39] N. Koblitz - p-adic variation of the zeta-function over families of varieties defined over finite fields. Compos. Math. 31 (1975), 119-218.

[40] S. Lang - *Complex Multiplication.* Grundl.255, Springer-Verlag (1983).

[41] S. Lang & H. Trotter - *Frobenius Distributions in GL_2-Extensions.* LNM 504, Springer - Verlag (1976).

[42] M. Larsen & R. Pink - On ℓ-independence of algebraic monodromy groups in compatible systems of representations. Invent. Math. 107 (1992), 603-636.

[43] M. Lazard - *Commutative Formal Groups.* LNM 443, Springer-Verlag, (1975).

[44] H. W. Lenstra & F. Oort - Simple abelian varieties having a prescribed formal isogeny type. Journ. Pure Appl. Algebra 4 (1974), 47-53.

[45] Ke-Zheng Li - Classification of supersingular abelian varieties. Math. Ann. 283 (1989), 333-351.

[46] Ke-Zheng Li - Actions of group schemes (I). Compos. Math. 80 (1991), 55-74.

[47] J. Lubin - Canonical subgroups of formal groups. Transact. AMS 251 (1979), 103-127.

[48] Yu. I. Manin - The theory of commutative formal groups over fields of finite characteristic. Usp. Math. 18 (1963), 3-90; Russ. Math. Surveys 18 (1963), 1-80.

[49] B. Mazur - Frobenius and the Hodge filtration, Bulletin of AMS 78 (1972), 653-667.

[50] W. Messing - *The Crystals Associated to Barsotti-Tate Groups: with Applications to Abelian Schemes.* LNM 264, Springer-Verlag (1972).

[51] L. Moret-Bailly - Familles de courbes et de variétés abéliennes sur $\mathbb{P}^1$. Sém. sur les pinceaux de courbes de genre au moins deux (L. Szpiro). Astérisque 86 (1981), 109-140.

[52] D. Mumford - A note of Shimura's paper "Discontinuous groups and abelian varieties". Math. Ann. 181 (1969), 345-351.

[53] D. Mumford - The structure of the moduli spaces of curves and abelian varieties. In: *Actes, Congrès international math.* (1970), tome 1, 457-465. Paris: Gauthier-Villars (1971).

[54] D. Mumford - *Lectures on Curves on an Algebraic Surface*. Annals of Math. Studies 59, Princeton U. Press, Princeton (1966).

[55] D. Mumford - *Abelian varieties*. Tata Inst. Fund. Res. & Oxford Univ. Press (1970) (2nd print. 1974).

[56] D. Mumford & J. Fogarty - *Geometric Invariant Theory*. 2nd ed., Springer-Verlag, Berlin-Heidelberg-New York (1982).

[57] R. Noot - Hodge classes, Tate classes, and local moduli of abelian varieties. Ph.D. thesis, Utrecht (1992).

[58] P. Norman & F. Oort - Moduli of abelian varieties. Ann. Math. 112 (1980), 413-439.

[59] N. Nygaard - On supersingular abelian varieties, In: *Algebraic Geometry, Proceedings, Ann Arbor (1981)* (ed. I. Dolgachev), LNM 1008, Springer-Verlag (1983), 83-101.

[60] T. Oda: The first De Rham cohomology groups and Dieudonné modules, Ann. Sci. Ecole Norm. Sup. 4e serie, t. 2 (1969), 63-135.

[61] T. Oda & F. Oort - Supersingular abelian varieties. Intl. Symp. on Algebraic Geometry, Kyoto (1977) (Ed. M.Nagata), Kinokuniya Book-store (1978), 595-621.

[62] A. Ogus - Supersingular K3 crystals. Journées Géom. Algébr. Rennes 1978, Vol. II. Astérisque 64, Soc. Math. France (1979), 3-86.

[63] A. Ogus - Hodge cycles and crystalline cohomology. In:*Hodge Cycles, Motives, and Shimura Varieties* (P. Deligne, J. Milne, A. Ogus, K-Y. Shih), LNM 900, Springer-Verlag (1982).

[64] F. Oort - *Commutative Group Schemes*. LNM 15, Springer-Verlag (1966).

[65] F. Oort - Lifting an endomorphism of an elliptic curve to characteristic zero. Indag. Math. 35 (1973), 466-470.

[66] F. Oort - The isogeny class of a CM-type abelian variety is defined over a finite extension of the prime field. Journ. Pure Appl. Algebra 3 (1973), 399-408.

[67] F. Oort - Subvarieties of moduli spaces. Invent. Math. 24 (1974), 95-119.

[68] F. Oort - Good and stable reduction of abelian varieties. Manuscr. Math. 11 (1974), 171-197.

[69] F. Oort - Which abelian surfaces are products of elliptic curves? Math. Ann. 214 (1975), 35-47.

[70] F. Oort - Isogenies of formal groups. Indag. Math. 37 (1975), 391-400.

[71] F. Oort - Lifting algebraic curves, abelian varieties and their endomorphisms to characteristic zero. Algebr. Geom. Bowdoin 1985. Proc. Symp. Pure Math. 46 (1987), AMS 1987; Part 2, 165-195.

[72] F. Oort - Endomorphism algebras of abelian varieties. In: *Algebr. Geom.*

and Commut. Algebra in Honor of M. Nagata (Ed. H. Hijikata et al.), Kinokuniya Co. (1988), Vol. II, 469-502.

[73] F. Oort - Hyperelliptic supersingular curves. In: *Arithmetic algebraic Geometry* (Eds.: G. van der Geer, F. Oort, J. Steenbrink), Progr. Math. 89, Birkhäuser, (1991), 247-284.

[74] F. Oort - CM-liftings of abelian varieties. J. Algebr. Geom. 1 (1992), 131-146.

[75] F. Oort - Moduli of abelian varieties and Newton polygons. C. R. Acad. Sci. Paris 312 (1991), 385-389.

[76] F. Oort - Moduli of abelian varieties in positive characteristic. In: *Barsotti Symposium in Algebraic Geometry (1991)* (Eds.: W. Messing, V. Cristante), Perspect. Math. 15, Acad. Press (1994), 253-276.

[77] F. Oort - Abelian varieties and Newton polygons. (To appear).

[78] F. Oort - Some questions in algebraic geometry. Manuscript, Univ. of Utrecht (1995), 22pp.

[79] F. Oort - A stratification of a moduli space of polarized abelian varieties in positive characteristic. Manuscript (1996), 15 pp. (To appear).

[80] M. Poletti - Differentiali esatti di prima specie su varietà abeliane. Ann. Scuola norm. sup. Pisa, Sci. fis. mat., 21 (1976), 107-110.

[81] H.-G. Quebbemann, W. Scharlau & M. Schulte - Quadratic and hermitian forms in additive and abelian categories. J. Algebra 59 (1979), 264-289.

[82] H.-G. Rück - Abelian surfaces and Jacobian varieties over finite fields. Compos. Math. 76 (1990), 351-366.

[83] W. Scharlau - *Quadratic and Hermitian Forms*. Grundl. 270, Springer-Verlag (1985).

[84] J-P. Serre - *Cohomologie galoisienne*. LNM 5, Springer-Verlag (1964).

[85] J-P. Serre - *Local Fields*. (Translation of *Corps locaux*), GTM 67, Springer-Verlag (1980).

[86] J-P. Serre - Complex multiplication. In: *Algebraic number theory* (Ed. J.W.S. Cassels & A.Fröhlich), Acad. Press (1976), 292-296. (Collected Papers J-P Serre II, 76).

[87] J-P. Serre - Groupes de Lie ℓ-adiques attachés aux courbes elliptiques. Colloque CNRS 143 (1966), 239-256 (see Œuvres, No. 70).

[88] J-P. Serre - *Lectures on the Mordell-Weil theorem*. Asp. Math. E15, Vieweg (1989).

[89] J-P. Serre & J. Tate - Good reduction of abelian varieties. Ann. Math. 88 (1968), 492-517. (Collected Papers J-P.Serre, II, 79).

[90] G. Shimura - On the theory of automorphic functions. Ann. Math. 70 (1959), 101-144.

[91] G. Shimura - On the zeta-function of an abelian variety with complex multiplication. Ann. Math. 94 (1971), 504-533.

[92] G. Shimura - Arithmetic of alternating forms and quaternion hermitian forms. J. Math. Soc. Japan 15 (1963), 33-65.

[93] G. Shimura - *Introduction to the Arithmetic Theory of Automorphic Functions.* Publ. Math. Soc. Japan 11, Iwanami Shoten & Princeton Univ. Press (1971).

[94] G. Shimura & Y. Taniyama - *Complex Multiplication of Abelian Varieties.* Publ. Math. Soc. Japan 6 (1961).

[95] T. Shioda - Supersingular K3 surfaces. In: *Algebraic Geometry*, Copenhagen 1978 (Ed. K. Lønsted). LNM 732, Springer-Verlag (1979), 564-591.

[96] J. H. Silverman - *The Arithmetic of Elliptic Curves.* GTM 106, Springer-Verlag (1986).

[97] J. Tate - Endomorphisms of abelian varieties over finite fields. Invent. Math. 2 (1966), 134-144.

[98] J. Tate - *Classes d'isogénie de variétés abéliennes sur un corps fini (d'après T.Honda).* Sém. Bourbaki Exp. 352 (1968/69). LNM 179, Springer-Verlag (1971).

[99] W. Tautz - Reduction of abelian varieties over number fields and supersingular primes. Ph.D. thesis, Queen's University, Kingston, Canada (1990).

[100] W. C. Waterhouse - Abelian varieties over finite fields. Ann. Sc. Ec. Norm. Sup. 2 (1969), 521-560.

[101] W.C. Waterhouse & J.S. Milne - Abelian varieties over finite fields, In: *Proc. Symp. Pure Math. vol. 20: Number Theory Inst. (Stony Brook)*, AMS (1971), 53-64.

[102] H. Weber - *Lehrbuch der Algebra*, Dritter Band: Elliptische Funktionen und algebraische Zahlen. Vieweg, Braunschwieg, 1908^2.

[103] N. Yui - Elliptic curves and canonical subgroups of formal groups. Journ. reine angew. Math. (Crelle), 303/304 (1978), 319-331.

[104] T. Zink - *Cartiertheorie kommutativer formaler Gruppen.* Teubner-Texte Math. 68, Teubner, Leipzig (1984).

[105] J. Milne - Abelian varieties. In: Arithmetic geometry (Editors: G. Cornell and J. Silverman). Springer-Verlag (1986), Chapter V, 103-150.

[106] R. Noot - Mumford's Shimura curves. Manuscript, September 1977, 47 pp.

Index

114

Springer
and the
environment

At Springer we firmly believe that an international science publisher has a special obligation to the environment, and our corporate policies consistently reflect this conviction.
We also expect our business partners – paper mills, printers, packaging manufacturers, etc. – to commit themselves to using materials and production processes that do not harm the environment. The paper in this book is made from low- or no-chlorine pulp and is acid free, in conformance with international standards for paper permanency.

 Springer

Printing: Weihert-Druck GmbH, Darmstadt
Binding: Buchbinderei Schäffer, Grünstadt

Lecture Notes in Mathematics

For information about Vols. 1–1485
please contact your bookseller or Springer-Verlag

Vol. 1526: J. Azéma, P. A. Meyer, M. Yor (Eds.), Séminaire de Probabilités XXVI. X, 633 pages. 1992.

Vol. 1527: M. I. Freidlin, J.-F. Le Gall, Ecole d'Eté de Probabilités de Saint-Flour XX – 1990. Editor: P. L. Hennequin. VIII, 244 pages. 1992.

Vol. 1528: G. Isac, Complementarity Problems. VI, 297 pages. 1992.

Vol. 1529: J. van Neerven, The Adjoint of a Semigroup of Linear Operators. X, 195 pages. 1992.

Vol. 1530: J. G. Heywood, K. Masuda, R. Rautmann, S. A. Solonnikov (Eds.), The Navier-Stokes Equations II – Theory and Numerical Methods. IX, 322 pages. 1992.

Vol. 1531: M. Stoer, Design of Survivable Networks. IV, 206 pages. 1992.

Vol. 1532: J. F. Colombeau, Multiplication of Distributions. X, 184 pages. 1992.

Vol. 1533: P. Jipsen, H. Rose, Varieties of Lattices. X, 162 pages. 1992.

Vol. 1534: C. Greither, Cyclic Galois Extensions of Commutative Rings. X, 145 pages. 1992.

Vol. 1535: A. B. Evans, Orthomorphism Graphs of Groups. VIII, 114 pages. 1992.

Vol. 1536: M. K. Kwong, A. Zettl, Norm Inequalities for Derivatives and Differences. VII, 150 pages. 1992.

Vol. 1537: P. Fitzpatrick, M. Martelli, J. Mawhin, R. Nussbaum, Topological Methods for Ordinary Differential Equations. Montecatini Terme, 1991. Editors: M. Furi, P. Zecca. VII, 218 pages. 1993.

Vol. 1538: P.-A. Meyer, Quantum Probability for Probabilists. X, 287 pages. 1993.

Vol. 1539: M. Coornaert, A. Papadopoulos, Symbolic Dynamics and Hyperbolic Groups. VIII, 138 pages. 1993.

Vol. 1540: H. Komatsu (Ed.), Functional Analysis and Related Topics, 1991. Proceedings. XXI, 413 pages. 1993.

Vol. 1541: D. A. Dawson, B. Maisonneuve, J. Spencer, Ecole d´ Eté de Probabilités de Saint-Flour XXI - 1991. Editor: P. L. Hennequin. VIII, 356 pages. 1993.

Vol. 1542: J.Fröhlich, Th.Kerler, Quantum Groups, Quantum Categories and Quantum Field Theory. VII, 431 pages. 1993.

Vol. 1543: A. L. Dontchev, T. Zolezzi, Well-Posed Optimization Problems. XII, 421 pages. 1993.

Vol. 1544: M.Schürmann, White Noise on Bialgebras. VII, 146 pages. 1993.

Vol. 1545: J. Morgan, K. O'Grady, Differential Topology of Complex Surfaces. VIII, 224 pages. 1993.

Vol. 1546: V. V. Kalashnikov, V. M. Zolotarev (Eds.), Stability Problems for Stochastic Models. Proceedings, 1991. VIII, 229 pages. 1993.

Vol. 1547: P. Harmand, D. Werner, W. Werner, M-ideals in Banach Spaces and Banach Algebras. VIII, 387 pages. 1993.

Vol. 1548: T. Urabe, Dynkin Graphs and Quadrilateral Singularities. VI, 233 pages. 1993.

Vol. 1549: G. Vainikko, Multidimensional Weakly Singular Integral Equations. XI, 159 pages. 1993.

Vol. 1550: A. A. Gonchar, E. B. Saff (Eds.), Methods of Approximation Theory in Complex Analysis and Mathematical Physics IV, 222 pages, 1993.

Vol. 1551: L. Arkeryd, P. L. Lions, P.A. Markowich, S.R. S. Varadhan. Nonequilibrium Problems in Many-Particle Systems. Montecatini, 1992. Editors: C. Cercignani, M. Pulvirenti. VII, 158 pages 1993.

Vol. 1552: J. Hilgert, K.-H. Neeb, Lie Semigroups and their Applications. XII, 315 pages. 1993.

Vol. 1553: J.-L- Colliot-Thélène, J. Kato, P. Vojta. Arithmetic Algebraic Geometry. Trento, 1991. Editor: E. Ballico. VII, 223 pages. 1993.

Vol. 1554: A. K. Lenstra, H. W. Lenstra, Jr. (Eds.), The Development of the Number Field Sieve. VIII, 131 pages. 1993.

Vol. 1555: O. Liess, Conical Refraction and Higher Microlocalization. X, 389 pages. 1993.

Vol. 1556: S. B. Kuksin, Nearly Integrable Infinite-Dimensional Hamiltonian Systems. XXVII, 101 pages. 1993.

Vol. 1557: J. Azéma, P. A. Meyer, M. Yor (Eds.), Séminaire de Probabilités XXVII. VI, 327 pages. 1993.

Vol. 1558: T. J. Bridges, J. E. Furter, Singularity Theory and Equivariant Symplectic Maps. VI, 226 pages. 1993.

Vol. 1559: V. G. Sprindžuk, Classical Diophantine Equations. XII, 228 pages. 1993.

Vol. 1560: T. Bartsch, Topological Methods for Variational Problems with Symmetries. X, 152 pages. 1993.

Vol. 1561: I. S. Molchanov, Limit Theorems for Unions of Random Closed Sets. X, 157 pages. 1993.

Vol. 1562: G. Harder, Eisensteinkohomologie und die Konstruktion gemischter Motive. XX, 184 pages. 1993.

Vol. 1563: E. Fabes, M. Fukushima, L. Gross, C. Kenig. M. Röckner, D. W. Stroock, Dirichlet Forms. Varenna, 1992. Editors: G. Dell'Antonio, U. Mosco. VII, 245 pages. 1993.

Vol. 1564: J. Jorgenson, S. Lang, Basic Analysis of Regularized Series and Products. IX, 122 pages. 1993.

Vol. 1565: L. Boutet de Monvel, C. De Concini, C. Procesi, P. Schapira, M. Vergne, D-modules, Representation Theory, and Quantum Groups. Venezia, 1992. Editors: G. Zampieri, A. D'Agnolo. VII, 217 pages. 1993.

Vol. 1566: B. Edixhoven, J.-H. Evertse (Eds.), Diophantine Approximation and Abelian Varieties. XIII, 127 pages. 1993.

Vol. 1567: R. L. Dobrushin, S. Kusuoka, Statistical Mechanics and Fractals. VII, 98 pages. 1993.

Vol. 1568: F. Weisz, Martingale Hardy Spaces and their Application in Fourier Analysis. VIII, 217 pages. 1994.

Vol. 1569: V. Totik, Weighted Approximation with Varying Weight. VI, 117 pages. 1994.

Vol. 1570: R. deLaubenfels, Existence Families, Functional Calculi and Evolution Equations. XV, 234 pages. 1994.

Vol. 1571: S. Yu. Pilyugin, The Space of Dynamical Systems with the C^0-Topology. X, 188 pages. 1994.

Vol. 1572: L. Göttsche, Hilbert Schemes of Zero-Dimensional Subschemes of Smooth Varieties. IX, 196 pages. 1994.

Vol. 1573: V. P. Havin, N. K. Nikolski (Eds.), Linear and Complex Analysis – Problem Book 3 – Part I. XXII, 489 pages. 1994.

Vol. 1574: V. P. Havin, N. K. Nikolski (Eds.), Linear and Complex Analysis – Problem Book 3 – Part II. XXII, 507 pages. 1994.

Vol. 1575: M. Mitrea, Clifford Wavelets, Singular Integrals, and Hardy Spaces. XI, 116 pages. 1994.

Vol. 1628: P.-H. Zieschang, An Algebraic Approach to Association Schemes. XII, 189 pages. 1996.

Vol. 1629: J. D. Moore, Lectures on Seiberg-Witten Invariants. VII, 105 pages. 1996.

Vol. 1630: D. Neuenschwander, Probabilities on the Heisenberg Group: Limit Theorems and Brownian Motion. VIII, 139 pages. 1996.

Vol. 1631: K. Nishioka, Mahler Functions and Transcendence.VIII, 185 pages.1996.

Vol. 1632: A. Kushkuley, Z. Balanov, Geometric Methods in Degree Theory for Equivariant Maps. VII, 136 pages. 1996.

Vol.1633: H. Aikawa, M. Essén, Potential Theory – Selected Topics. IX, 200 pages.1996.

Vol. 1634: J. Xu, Flat Covers of Modules. IX, 161 pages. 1996.

Vol. 1635: E. Hebey, Sobolev Spaces on Riemannian Manifolds. X, 116 pages. 1996.

Vol. 1636: M. A. Marshall, Spaces of Orderings and Abstract Real Spectra. VI, 190 pages. 1996.

Vol. 1637: B. Hunt, The Geometry of some special Arithmetic Quotients. XIII, 332 pages. 1996.

Vol. 1638: P. Vanhaecke, Integrable Systems in the realm of Algebraic Geometry. VIII, 218 pages. 1996.

Vol. 1639: K. Dekimpe, Almost-Bieberbach Groups: Affine and Polynomial Structures. X, 259 pages. 1996.

Vol. 1640: G. Boillat, C. M. Dafermos, P. D. Lax, T. P. Liu, Recent Mathematical Methods in Nonlinear Wave Propagation. Montecatini Terme, 1994. Editor: T. Ruggeri. VII, 142 pages. 1996.

Vol. 1641: P. Abramenko, Twin Buildings and Applications to S-Arithmetic Groups. IX, 123 pages. 1996.

Vol. 1642: M. Puschnigg, Asymptotic Cyclic Cohomology. XXII, 138 pages. 1996.

Vol. 1643: J. Richter-Gebert, Realization Spaces of Polytopes. XI, 187 pages. 1996.

Vol. 1644: A. Adler, S. Ramanan, Moduli of Abelian Varieties. VI, 196 pages. 1996.

Vol. 1645: H. W. Broer, G. B. Huitema, M. B. Sevryuk, Quasi-Periodic Motions in Families of Dynamical Systems. XI, 195 pages. 1996.

Vol. 1646: J.-P. Demailly, T. Peternell, G. Tian, A. N. Tyurin, Transcendental Methods in Algebraic Geometry. Cetraro, 1994. Editors: F. Catanese, C. Ciliberto. VII, 257 pages. 1996.

Vol. 1647: D. Dias, P. Le Barz, Configuration Spaces over Hilbert Schemes and Applications. VII, 143 pages. 1996.

Vol. 1648: R. Dobrushin, P. Groeneboom, M. Ledoux, Lectures on Probability Theory and Statistics. Editor: P. Bernard. VIII, 300 pages. 1996.

Vol. 1649: S. Kumar, G. Laumon, U. Stuhler, Vector Bundles on Curves – New Directions. Cetraro, 1995. Editor: M. S. Narasimhan. VII, 193 pages. 1997.

Vol. 1650: J. Wildeshaus, Realizations of Polylogarithms. XI, 343 pages. 1997.

Vol. 1651: M. Drmota, R. F. Tichy, Sequences, Discrepancies and Applications. XIII, 503 pages. 1997.

Vol. 1652: S. Todorcevic, Topics in Topology. VIII, 153 pages. 1997.

Vol. 1653: R. Benedetti, C. Petronio, Branched Standard Spines of 3-manifolds. VIII, 132 pages. 1997.

Vol. 1654: R. W. Ghrist, P. J. Holmes, M. C. Sullivan, Knots and Links in Three-Dimensional Flows. X, 208 pages. 1997.

Vol. 1655: J. Azéma, M. Emery, M. Yor (Eds.), Séminaire de Probabilités XXXI. VIII, 329 pages. 1997.

Vol. 1656: B. Biais, T. Björk, J. Cvitanić, N. El Karoui, E. Jouini, J. C. Rochet, Financial Mathematics. Bressanone, 1996. Editor: W. J. Runggaldier. VII, 316 pages. 1997.

Vol. 1657: H. Reimann, The semi-simple zeta function of quaternionic Shimura varieties. IX, 143 pages. 1997.

Vol. 1658: A. Pumariño, J. A. Rodríguez, Coexistence and Persistence of Strange Attractors. VIII, 195 pages. 1997.

Vol. 1659: V. Kozlov, V. Maz'ya, Theory of a Higher-Order Sturm-Liouville Equation. XI, 140 pages. 1997.

Vol. 1660: M. Bardi, M. G. Crandall, L. C. Evans, H. M. Soner, P. E. Souganidis, Viscosity Solutions and Applications. Montecatini Terme, 1995. Editors: I. Capuzzo Dolcetta, P. L. Lions. IX, 259 pages. 1997.

Vol. 1661: A. Tralle, J. Oprea, Symplectic Manifolds with no Kähler Structure. VIII, 207 pages. 1997.

Vol. 1662: J. W. Rutter, Spaces of Homotopy Self-Equivalences – A Survey. IX, 170 pages. 1997.

Vol. 1663: Y. E. Karpeshina; Perturbation Theory for the Schrödinger Operator with a Periodic Potential. VII, 352 pages. 1997.

Vol. 1664: M. Väth, Ideal Spaces. V, 146 pages. 1997.

Vol. 1665: E. Giné, G. R. Grimmett, L. Saloff-Coste, Lectures on Probability Theory and Statistics 1996. Editor: P. Bernard. X, 424 pages, 1997.

Vol. 1666: M. van der Put, M. F. Singer, Galois Theory of Difference Equations. VII, 179 pages. 1997.

Vol. 1667: J. M. F. Castillo, M. González, Three-space Problems in Banach Space Theory. XII, 267 pages. 1997.

Vol. 1668: D. B. Dix, Large-Time Behavior of Solutions of Linear Dispersive Equations. XIV, 203 pages. 1997.

Vol. 1669: U. Kaiser, Link Theory in Manifolds. XIV, 167 pages. 1997.

Vol. 1670: J. W. Neuberger, Sobolev Gradients and Differential Equations. VIII, 150 pages. 1997.

Vol. 1671: S. Bouc, Green Functors and G-sets. VII, 342 pages. 1997.

Vol. 1673: F. D. Grosshans, Algebraic Homogeneous Spaces and Invariant Theory. VI, 148 pages. 1997.

Vol. 1674: G. Klaas, C. R. Leedham-Green, W. Plesken, Linear Pro-p-Groups of Finite Width. VIII, 115 pages. 1997.

Vol. 1676: P. Cembranos, J. Mendoza, Banach Spaces of Vector-Valued Functions. VIII, 118 pages. 1997.

Vol. 1677: N. Proskurin, Cubic Metaplectic Forms and Theta Functions. VIII, 196 pages. 1998.

Vol. 1678: O. Krupková, The Geometry of Ordinary Variational Equations. X, 251 pages. 1997.

Vol. 1679: K.-G. Grosse-Erdmann, The Blocking Technique. Weighted Mean Operators and Hardy's Inequality. IX, 114 pages. 1998.

Vol. 1680: K.-Z. Li, F. Oort, Moduli of Supersingular Abelian Varieties. V, 116 pages. 1998.

General Remarks

Lecture Notes are printed by photo-offset from the master-copy delivered in camera-ready form by the authors. For this purpose Springer-Verlag provides technical instructions for the preparation of manuscripts.

Careful preparation of manuscripts will help keep production time short and ensure a satisfactory appearance of the finished book. The actual production of a Lecture Notes volume normally takes approximately 8 weeks.

Authors receive 50 free copies of their book. No royalty is paid on Lecture Notes volumes.

Authors are entitled to purchase further copies of their book and other Springer mathematics books for their personal use, at a discount of 33,3 % directly from Springer-Verlag.

Commitment to publish is made by letter of intent rather than by signing a formal contract. Springer-Verlag secures the copyright for each volume.

Addresses:

Professor A. Dold
Mathematisches Institut
Universität Heidelberg
Im Neuenheimer Feld 288
D-69120 Heidelberg
Federal Republic of Germany

Professor F. Takens
Mathematisch Instituut
Rijksuniversiteit Groningen
Postbus 800
NL-9700 AV Groningen
The Netherlands

Springer-Verlag, Mathematics Editorial
Tiergartenstr. 17
D-69121 Heidelberg
Federal Republic of Germany
Tel.: *49 (6221) 487-410